Measurement of
HUMAN
LOCOMOTION

Measurement of
HUMAN
LOCOMOTION

Vladimir Medved, Ph.D.

CRC Press
Taylor & Francis Group
Boca Raton London New York

CRC Press is an imprint of the
Taylor & Francis Group, an **informa** business

CRC Press
Taylor & Francis Group
6000 Broken Sound Parkway NW, Suite 300
Boca Raton, FL 33487-2742

First issued in paperback 2019

© 2001 by Taylor & Francis Group, LLC
CRC Press is an imprint of Taylor & Francis Group, an Informa business

No claim to original U.S. Government works

ISBN-13: 978-0-8493-7675-7 (hbk)
ISBN-13: 978-0-367-39777-7 (pbk)
Library of Congress Card Number 00-048636

Library of Congress Cataloging-in-Publication Data

Medved, Vladimir.
    Measurement of human locomotion / Vladimir Medved.
       p. cm.
    Includes bibliographical references and index.
    ISBN 0-8493-7675-0 (alk. paper)
      1. Human locomotion—Measurement. 2. Movement disorders—Diagnosis. I. Title.

    QP301 .M384 2000
    612.7′6′0287—dc21
                                                         00-048636

Visit the Taylor & Francis Web site at
http://www.taylorandfrancis.com

and the CRC Press Web site at
http://www.crcpress.com

To my wife Vesna

and my children, Ranko and Sara

Preface

The importance of measurements to properly assess human locomotion is increasingly recognized. Already well established as an experimental scientific research tool, human locomotion measurements are frequently a routine clinical application. Fields of application encompass both healthy and pathological locomotion encountered in rehabilitation medicine, orthopedics, kinesiology, sports science, and other related fields.

This volume provides comprehensive description of instrument systems for measurement of kinematics of human movement, kinetic quantities experienced by the human body in contact with the ground, and myoelectric changes associated with locomotor activity. Physical principles governing the operation of several measurement devices and relevant mathematics and engineering are presented, as well as signal processing issues that must be addressed in order to obtain and use quantitative measurement variables in the biomechanical context. Measurement data acquisition, processing, and presentation to the user in a computer-based laboratory environment are explained. The ultimate goal is to contribute to the diagnostics and treatment of specific locomotion patterns. References to major historical landmarks in the development of measurement methodology are provided as well. Selected experimental data are shown and interpreted to illustrate the methods, some originating from the author's own research. Consequently, the reader will gain insight into the working principles, typical uses, and comparative advantages of a number of instruments, such as simple electrogoniometers, sophisticated stereometric instruments to capture human body kinematics, imbedded force plates, distributed pressure measurement systems, wire and telemetry electromyographs, etc.

Systems oriented and interdisciplinary in character, this volume addresses biomedical engineers, active in industry or the clinical environment, physicians, kinesiologists, physical therapists, and students and researchers of human movement in clinics and academia. By focusing on locomotion measurements, the volume attempts to complement classical biomechanics, neurophysiology, and motor control-oriented texts.

Author

Vladimir Medved, Ph.D., is Associate Professor of Biomechanics, Faculty of Physical Education, University of Zagreb, in Zagreb, Croatia. He also teaches elective courses at the Faculty of Electrical Engineering and Computing.

Dr. Medved received his B.Sc., M.Sc., and Ph.D. degrees in electrical engineering in 1974, 1977, and 1988, respectively, from the University of Zagreb, Faculty of Electrical Engineering (now known as the Faculty of Electrical Engineering and Computing). He was a Research Engineer from 1977 to 1982 at the Institute for Electronics, Telecommunications, and Automation of Radioindustry in Zagreb, developing mobile radiocommunication systems and microprocessor applications. Since 1982 he has been a member of the Faculty of Physical Education at the University of Zagreb, developing and directing the Biomechanics Laboratory facility and researching biomechanics and collaborating in exercise physiology, sports medicine, and anthropology. He became Assistant Professor in 1992 and Associate Professor in 1996.

Dr. Medved's fields of special interest are biomedical engineering, biomechanics, electrophysiological signal measurement and processing, kinesiological electromyography, and measurement of locomotion in particular. He has published, as author or co-author, in archive journals and has participated in international conferences. In 1979/1980, he was at Harvard University and the Massachusetts Institute of Technology as a Baloković Scholar; in 1990/1991, he was at Harvard University as a Postdoctoral Fulbright Scholar; and in 1984 and 1993, he was at Uppsala University in Sweden. Dr. Medved has led several research projects and presently conducts research in neuromuscular biomechanical diagnostics of healthy and pathological locomotion, sponsored by the Croatian Ministry of Science and Technology. He serves as one of the editors for the journal *Kinesiology*, published in Zagreb, and is a Member of the Council of the *Croatian Sports Medical Journal* (in Croatian). He has also been a member of the organizing committees of international scientific symposia.

Dr. Medved is a member of the Zagreb Electrotechnical Society, the Croatian Medical and Biological Engineering Society, the International Federation of Medical and Biological Engineering (IFMBE), the International Society of Electrophysiology and Kinesiology (ISEK), a collaborating member of the Croatian Academy of Engineering, and a member of the New York Academy of Sciences.

Acknowledgment

I am grateful to CRC Press for accepting my manuscript for this book and for having enough patience and flexibility to tolerate my delays in finishing the project. While there is obviously great professionalism behind the operation of this publishing house, there are individuals with whom I had direct contact, each in their respective capacity, who proved decisive for the accomplishment of this book and to whom I owe my special thanks: Marsha Baker, Liz Covello, Barbara Norwitz, Carol Hollander, Carolyn Lea, Pat Roberson, and Jonathan Pennell. I wish to thank them particularly; it has been a pleasure to work with them.

I also thank BTS Bioengineering Technologies and Systems (ELITE),... Selcom (SELSPOT), Northern Digital (OPTOTRAK), AMTI, BLH Electronics, Kistler, and Oxford Metrics (VICON) for the courtesy of allowing me use of their commercial and technical material in some of the figures.

Table of Contents

1 Introduction

Human locomotion may be defined as the action by which the body as a whole moves through aerial, aquatic, or terrestrial space. Locomotion is achieved by coordinated movements of the body segments, taking advantage of an interaction of internal and external forces (Cappozzo et al. 1976, Reference 1), and is accomplished through the action of the neuro-musculo-skeletal system. In both healthy and pathological locomotion, the fact that it is possible to take measurements is of great significance. For example, various effects and manifestations of locomotion that either directly or indirectly mirror the function of the neuro-musculo-skeletal system may be measured. Three distinct subsets of physical variables are included when measuring locomotion: kinematic data, which describe movement geometry, forces and moments that are exerted when the body and its surroundings interact, the so-called kinetic or dynamic data, and bioelectric changes associated with skeletal muscle activity, the so-called myoelectric, i.e., electromyographic (EMG) signals. Taken together, these data provide a comprehensive picture of this phenomenon.

What is the purpose of these measurements? In various pathologies of the locomotor apparatus and gait disturbances, for instance, the purpose might be to contribute to more appropriate diagnostics and treatment, i.e., therapy. For example, measurement data reveal those features of this phenomenon that are not accessible via mere visual observation and other medical/clinical methods. The most prominent field of application in this respect is probably that concerned with orthotic and prosthetic devices for extremities used in pathologies and traumas of the locomotor system, i.e., in medical rehabilitation. Such issues have most certainly contributed significantly to the motivation for measuring movement structures.

Why, however, should one measure healthy locomotion? One area of research encompasses the broad spectrum of sports activities. Data obtained by measuring structures in sporting movement may be important from the standpoint of acquiring proper technique, correction of errors in technique, optimization of the training process, etc. Answers to several questions may be sought. Are the gymnast's body rotations during the airborne phase of the somersault smooth and energetically efficient? Is the activity of the leg muscles of an athlete explosive enough when leaving the ground during a high jump? What moments of force develop in the hip joint during running? In what manner do modern athletic footwear modify foot loading during contact with the floor?

Ergonomics, i.e., man-machine interaction, is also an area that may benefit from measuring movement structures. There are a multitude of working situations where it is of interest to estimate quantitatively the loading pattern induced by certain dynamic actions or static body positions and, in connection with this, the organism's energy expenditure. These kinds of procedures might provide a basis for improvement of the working process and simultaneously decrease chronic, potentially

1

traumatic actions on the body. Finally, concerning bionics, human movement might represent a model for designing locomotion automata and robots. Therefore, the measurement of skilled and virtuosic human movements might provide relevant information in this field. In research laboratories around the world, work is being done in a highly interdisciplinary spirit, incorporating biology and engineering. Physiology, biomechanics, kinesiology, robotics, ergonomics, neuroscience, and artificial intelligence all merge in this endeavor. By using computer simulations of locomotion and comparing the results obtained with measurement data, the goal is to solve problems such as the design of artificial skeletal muscles, the construction of mobile robots, telerobotic control, the construction of intelligent prostheses, etc. These issues may be relevant to biomedical, military, and consumer industries.

1.1 ABOUT THE BOOK

This book describes the methods, technical devices, and procedures used when measuring both pathological and/or healthy human locomotion (aquatic movements are not considered). These engineering solutions, systems, and procedures facilitate a more objective evaluation and a better understanding of the locomotor function, which is yet to be fully understood. One may acquire new and better insight into the mechanisms of action and function of the neuro-musclo-skeletal system, a physiological "creation" which—if compared to engineering—might be regarded as being one of the most complex automatic control systems in the natural world.

Chapter 1 is a review of the major historical landmarks in the development of locomotion measurement methods. Special attention is primarily paid to kinematic variables. Historical insights into certain more recent measurement methods of kinetic, and particularly, myoelectric variables are given in more detail in Chapters 5 and 6, respectively.

Chapter 2 provides an answer to the question: which variables have to be measured and why? The methodology of human movement study is presented, which provides the source of required measurement methods. On one hand, measurements are determined by biomechanical modeling of the human body, thereby enabling quantitative characterization of locomotion by treating the body as a complex multisegmental mechanical system. On the other, basic neurophysiological mechanisms of the locomotor apparatus are briefly presented, giving insight into the biocommunication and bioenergetic processes vital for the realization of movement structures. A summary of skeletal muscle biomechanics is found at the end of this chapter.

Chapter 3, focusing on what is common to all measurement methods, begins with a presentation of the global structure of the measurement system. The analog-to-digital conversion procedure, enabling the interface of the analog measurement data to the digital computer, is also described. Overall requirements of locomotion measurement systems from the users' standpoint are summarized at the end of this chapter.

Measurement methods and procedures fall into three categories: kinematic (Chapter 4), kinetic (Chapter 5), and myoelectric (Chapter 6). While kinematic and kinetic locomotion variables are mechanical entities, myoelectric signals are physiological variables, originating in the human body. Therefore, in Chapter 6, a description of myoelectric signal genesis is also given. Each particular measurement method, i.e., system, is described through the basic physical and/or engineering working principles. The essential parts of the engineering realization of systems are explained at the level of block schematics or, when necessary, electronic circuit or mathematical algorithms. Measurement errors are evaluated, from those caused by sensor features (accuracy, linearity, frequency characteristics, possible hysteresis, durability, etc.) to those appearing in the information processing chain, before the final result is presented to the user. Particular methods (or groups of methods, providing they measure the same variables using physically or technically different procedures) are illustrated by typical results with an interpretation, including the author's own results, when appropriate.

Chapter 7 provides several examples of how comprehensive systems which integrate kinematic, kinetic, and EMG measurements, and are supported by specific data processing and interpretation facilities, are applied to measuring, analyzing, and diagnosing locomotion. The emphasis is on practical clinical applications and standardization of methods. A noninvasive, automated kinematic measurement method, currently under development, is presented. Figure 1.1 illustrates a subject being measured using three groups of measurement variables, which monitor his locomotion comprehensively.

The measurement of locomotion is viewed in a broader sense. That is, detection, acquisition, and collection of respective quantitative data of the aforementioned

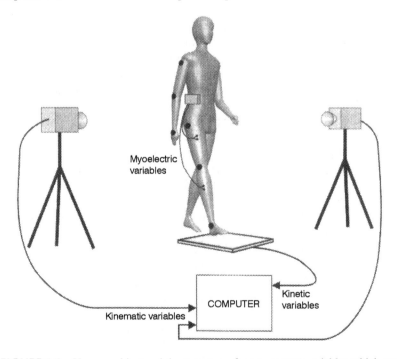

FIGURE 1.1   Human subject and three groups of measurement variables which monitor his locomotion comprehensively.

physical variables for describing human movement are included, as well as their subsequent processing and interpretation. The goal is to provide objectivized quantitative evaluation and diagnostics of the locomotion measured. Pierrynowski[2] describes the aim of laboratory measurement procedures in this area well: "Essentially, locomotion laboratories identify the location of a subject on a body movement functionality scale at a particular time. This scale ranges from complete motor disability (death) to elite athletes, with coma, amputation, paralysis, normal, and athletic in-between. The clinical or coaching teams then attempt to shift the patient or athlete along this continuum."

Human locomotion measurements are recognized as an experimental scientific research tool. Practical diagnostic possibilities of corresponding methods in direct clinical application and testing, however, are sometimes regarded with skepticism.[3] According to this viewpoint, in spite of the application of sophisticated technology, developments in this field of research did not give us valid, reliable, and feasible clinical diagnostic and evaluation procedures comparable to those of some other systems of the body such as the cardiovascular, pulmonary, or nervous systems. Messenger,[4] for instance, gave an overview of the clinical application of gait analysis—a basic subfield of locomotion measurement—in Great Britain at the beginning of the 1990s. A typical laboratory facility was equipped with a force platform and a pedobarographic system (kinetic quantities), a video system (kinematic quantities), and, in 50% of cases, an EMG recording instrument. On average, ten subjects were measured per month in the laboratory (which might indicate that corresponding procedures were not completely standardized, nor time efficient, and hence not accepted in clinical practice). However, mostly due to the significant technological development which has taken place recently, opinions have become noticeably more optimistic, primarily regarding the application of locomotion measurements in certain specific, previously diagnosed pathologies. Consider, for instance, the case of individuals with cerebral palsy, whose gait and motorics are severely disturbed. According to medical clinicians, in individuals with various specific forms of this syndrome, measuring locomotion might contribute significantly to overall differential diagnostics, prognostics, adequate (surgical) treatment, and the evaluation of the treatment, as well as rehabilitation follow-up of the locomotor apparatus.[5–12] Furthermore, clinical application of locomotion measurement can be an invaluable asset when evaluating prostheses and orthoses for the lower extremities, as well as when conducting a variety of sports testing procedures. These are the issues discussed in Chapter 7. Certain examples from previous chapters are also relevant.

The world market is flooded with books dealing with biomechanics, neurophysiology, and the motorics of human gait and locomotion, primarily the functioning of the locomotor apparatus. Although a great number of books also discuss certain medical instrumentation, as well as instrumentation in bordering fields, particular locomotion measurement procedures are usually described in papers in specialized journals, in conference proceedings, or in addenda to medical and kinesiological books. Comprehensive volumes targeted primarily at locomotion measurement systems are rare. This book is an attempt in this direction.

1.2   THE HISTORY OF LOCOMOTION MEASUREMENT

A short survey of the historical development in the field of human locomotion measurement and analysis follows, based on the references.[1,3,9,13-32]

Aristotle (384-322 B.C.), the famous ancient Greek philosopher, analyzed animal movement qualitatively in 344 B.C. in his book "De motu animalium." He wrote about animal locomotion, attempting to analyze the phenomenon geometrically. He designed a model of the animated mechanics of the extremities according to the number and type of joints and distinguished particular parts of an animal's body according to their functions. Careful observation was his sole "measuring instrument." He was the first to describe the action of muscles and movement in the joints during locomotion. He compared the complex control of movement in a biological organism with control within a city-state.

In connection with locomotion, albeit not directly with its measurement, Galen (131-201) is also worth mentioning. He was a physician to gladiators in ancient Greece and the term myology ("De motu musculorum") is linked with his name.

For a long time thereafter, there were no written documents in the field of human locomotion. However, Gruner mentions Avicenna (980-1037) ("A Treatise on the Canon of Medicine of Avicenna," London, 1930, according to Reference 27).

Leonardo da Vinci (1452-1519), an Italian painter, sculptor, builder, and scientist, attempted to understand and explain the phenomenon of movement. Interested in anatomy, he created precise drawings of human skeletal muscles and the way they were attached to the skeleton based on postmortem dissections. His drawings of the human body, its proportions, organs, and functions, are considered to be among the highest achievements in Renaissance science. Some of his observations are presented in: *On the Human Body* (O'Malley and Saunders, Henry Schuman, New York, 1952).

Andreas Vesalius (1514-1564), a Belgian anatomist, followed in the footsteps of Galen and later made corrections to his work. Weber and Weber (1836, according to Reference 28) also mentioned Fabricius ab Acquapedente (1618).

The works of Galileo Galilei (1564-1642), an Italian physicist, astronomer, and mathematician, and Isaac Newton (1642-1727), an English mathematician, physicist, astronomer, and philosopher, laid the theoretical and experimental foundations for movement analysis based on physical principles. Newton's work, "Philosophiae Naturalis Principia Mathematica," dates back to 1686. Based on this work, all motion in the universe could be described and/or predicted as long as this motion was at a relative speed slower than the speed of light.

The measurement and analysis of locomotion in a more objective and quantitative manner began, however, with Galileo's pupil, the Renaissance scientist Giovanni Alfonso Borelli (1608-1679). Borelli, an Italian physiologist and astronomer, was the main representative of iatromechanics, a medical theory which viewed the human organism as a machine, reducing physiological and pathological phenomena to mechanical phenomena. Another important iatromechanist was Đuro Armeno Baglivi, 1668-1707, born in Dubrovnik. Borelli ("De motu animalium ex principio mechanico statico," 1680) was the first to apply Galileo's scientific method to the phenomenon of movement and therefore may be considered to be the initiator of bio-dynamics and biokinematics of locomotion and hence the founder of biomechanics. He viewed bones as mechanical levers moved by muscles according to mathematical principles. He gave suggestions on how to determine mechanical forces influencing the biological system from a static point of view. By integrating his knowledge of mathematics, physics, and anatomy, Borelli was the first to determine the center of the human body's mass by balancing the body around a prismatic pivot within three mutually orthogonal planes. He performed an interesting experiment (using himself) in 1679 by putting two vertical posts in the intended movement direction. On walking toward them, he failed to keep them in the line of sight, which indicated a lateral sway during gait.[16] Gassendi (1592-1655) also reported on a similar experiment consisting of walking along a wall with an extended hand.

The Dutchman, Hermann Boerhaave (1668-1738), a physician, chemist, and botanist, followed the work of Newton and was the first to consider dynamics, i.e., the kinetics of movement, taking inertial influences into account. In 1703 he gave a speech at the University of Leiden entitled "On the Use of Mechanical Method in Medicine" in which he foresaw future events in the field of biomechanics that would take place 200 years later with the development of data acquisition and processing systems.

Leonhard Euler (1707-1783), a mathematician, introduced differential equations for the description of rigid body movement which were important for the development of kinematics.[24]

In 1798, the French physician P. I. Bathrez (1734-1806) presented a more comprehensive theory of movement in men and animals entitled "Nouvelle mechanique des movements de l'homme et des animaux."

The discovery of bioelectricity by Luigi Galvani (1737-1798), an Italian physicist and physician, in 1791,[32] is significant in the history of both electrophysiology and monitoring skeletal muscle functions, which are important for registering and studying locomotion. The practical application of this discovery in the study of locomotion, however, took place at the beginning of 20th century with the development of a technique for registering bioelectric phenomena. This will be discussed in more detail in Chapter 6.

The Weber brothers, Wilhelm (1804-1891), a physicist, and Eduard (1806-1871), a physician, from Göttingen, conducted physical-physiological studies of gait, using the observation techniques of the time: chronograph, meter, and an optical instrument, the diopter. They were the first to publish a scientific treatise on gait, measuring it systematically using optical means, prior to the discovery of photography (Weber, W. and E., "Die Mechanik der menschlichen Gehwerkzeuge. Eine anatomisch-physiologische Untersuchung," Göttingen, 1836). They used chronographs, developed in the second half of the 17th century, which made possible measurement of elapsed time in (large) parts of a second. In this way, it was possible to calculate stride length and gait velocity. They were the first to study the support phase and swing phase in gait, as well as the relationship between time and length of stride. Having represented the swinging leg as an inverted double pendulum, they conducted numerical calculations for this model. This "pendulum theory" for the swinging leg (Van Hussen, 1973, according to Reference 28), the idea that swinging occurs entirely passively, although refuted later, was the first attempt at mathematically modeling the aspects of human movement. In trying to explain the laws of human movement, they considered "the principle of least muscular effort" to be fundamental. It is interesting to note that Wilhelm Weber, famous for the science of wave propagation (after whom the magnetic flux measurement unit is named), spent his school days conducting experiments with the assistance of his second brother, the older Ernst Heinrich (1795-1878), a physiologist, anatomist, and psychologist (Weber-Fechner law).

The discovery of photography by Louis Jacques Mandé Daguerre (1787-1852), a French painter, in 1839 was an epochal event. The first photographic method is named after Daguerre—daguerreotypy. Discovery, known as diorama—a partially transparent image induced by reflected and emitted light to create an illusion of change—is the precedent of cinematography. Another important event in the history of photography was the discovery of the single flash photographic exposure by William Henry Talbot (1800-1877), an English physicist and chemist, in 1859. Measurement of locomotion in the modern sense, however, is due to Eadweard Muybridge (1830-1904), a British photographer (real name Muggeridge), who, working in the U.S., marked the beginning of the study of locomotion as a dynamic natural phenomenon by applying photography, thereby leading to the quantitative approach. Muybridge's major preoccupation was photographing geographic landscapes in California, such as Yosemite Park. Commissioned and financially supported by Leland Stanford (an ex-governor of California at the time and founder of Stanford University, who also loved racing horses), Muybridge started using photography to record horses running in Sacramento, California. It was partly a result of a wager. The matter to be settled was whether or not a horse left the ground completely at any point in time while running. The project took place from 1872 to 1877. Muybridge succeeded in developing a photographic emulsion capable of recording 1/1000 (1/2000 according to some sources) of a second. To measure the horse's stride, he used a series of cameras—12 and later 24. They were activated with delays, with the aid of a so-called "rotating commutator mechanism for magnetic shutter release." As a result, he obtained a time series of pictures of the moving animal on a wet glass plate. The analysis of these pictures gave the first real sequence of the animal's movements ("The Horse in Motion," 1882). The wager was resolved when it was found that there did indeed exist time intervals during which the horse left the ground completely. This was the basis of modern cinematography which appeared in 1895. In Palo Alto, California, and later, at the University of Pennsylvania, where he achieved the speed of 1/6000 of a second, Muybridge recorded about 100,000 plates of animal and human locomotion. Muybridge showed detailed sequences of various human locomotor activities recorded from three angles (see *The Human Figure in Motion,* Dover, New York, 1955). In 1887 he published 11 volumes of photographs entitled *Animal Locomotion.* Despite criticism concerning the interpretation of the recorded photographs (Braun, 1993, according to Reference 23), there is no doubt that

Muybridge's work offered a new "kinematic scientific language." His insights and discoveries also significantly influenced the art of painting.

Francis X. Dercum ("The Walk in Health and Disease," in *Trans. College Phys.,* 10, pp. 308-338), a neurologist at the University of Pennsylvania, used Muybridge's achievement in his clinical work. He recognized the importance of photography for the determination of gait kinematic features which were inaccessible by direct visual observation, e.g., in the study of ataxia, in which muscle coordination is disturbed resulting in stereotypical movement patterns, such as in gait (a disturbance in the functioning of the cerebellum). Dercum showed displacement curves for normal and pathological gait as a function of time, derived from Muybridge's recordings.

Étienne-Jules Marey (1830-1904), a French physician and physiologist, and his pupils, M. G. Carlet ("Essai expérimentale sur la locomotion humaine—étude de la marche," *Ann. Sci. Nat. Zool.,* 16 (Séries 5, art. 6), pp. 1–92, 1872) and Vierordt (1881), also applied photography. Marey began research into human and animal loco-motion in Paris around 1870. He was one of the most acclaimed physiologists of his time and published extensively on the subjects of blood circulation, muscular con-traction, respiration, human and animal locomotion, and bird and insect flight. He researched the marching of soldiers. His book ("La méthode graphique dans les sci-ences expérimentales et particuliérement en physiologie et en médicine," Paris, Masson, 1872, 2nd edition with addendum, "Le dévelopment de la méthode graphique par l'emploi de la photographie," 1885) is a fundamental work in experi-mental methodology and instrumentation. He invented several devices for registering the action of human organs graphically. Marey was the first to develop a *de facto* bio-mechanical laboratory facility: the Station Physiologique in the Park des Princes (today the site of Roland Gaross tennis courts) where a horizontal circular track was set up, 500 m in circumference and equipped with measuring instruments. According to Marey,[21] "Registering dynamometers, spirometers, pedometers, and various appa-ratus for the measurement of objects under observation are devoted to the study of human locomotion. In addition, pneumographs, sphygmographs, and cardiographs enable the investigator to study the effect of athletic exercises on the function of organic life, and to follow step by step the improvement under training." This part of Marey's description of the Station Physiologique demonstrates that it was a precur-sor of modern biomechanical-physiological laboratory facilities. By arguing the need for such a facility, Marey historically added to the achievements of Galileo, Borelli, and the Weber brothers, claiming that it was necessary to introduce greater accuracy when studying locomotion. His important technical solution for the registration of variable processes, including the locomotor process, is known as the graphic method which provided registration of the detected time change of certain variables (physi-cal, physiological) via a pneumatic transportation tube and a stylus. The name given to this device, consisting of a rotating drum and a stylus, was the odograph.

During one period of their research, prior to the application of photography, Marey and Carlet took locomotion measurements by using the above-mentioned graphic method and specially constructed measurement devices. Marey presented this basic pneumatic principle of detection of foot-floor contact. The sole was com-posed of a thick sheet of rubber with a hollow chamber. This cavity communicated with the recording tambour through a long flexible tube. In this manner, a change in pressure in the chamber was registered. Carlet (1872) applied the method to register step sequence. Measurements took place on a circular path, 20 m in diameter. The subject was required to carry a fairly massive mechanism serving to transmit measurement change to the registration site. Marey modified the method, making the recorder portable (1873). This was a kymograph—a slowly rotating drum covered with smoked paper (in principle, the description fits the above-mentioned odograph). Figure 1.2 shows this witty, but impractical method. According to Harry,[18] the kymograph was originally developed in 1847 by Carl Ludwig (1816-1895). It revolutionized the field of physiology by producing the first permanent record of a physiological process—the change in blood pressure. The drum revolved at a constant speed while the stylus registered any change detected. This method, however, could not continuously register movement trajectories of one or more points on the body.

The German physiologist Karl Hermann Vierordt (1881) described an ingenious method using thin ink jets which sprayed ink from styluses attached vertically to the subject's shoes or joints; these jets hit long paper strips placed on the ground and hung alongside the path the subject would take (Figure 1.3). In addition, small cotton wads soaked in ink and attached to the shoes marked shoe position during

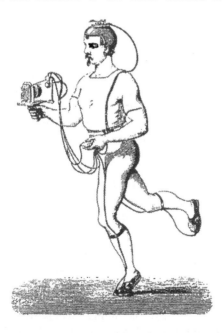

FIGURE 1.2 Runner from the 1870s carrying a clockwork recorder (kymograph) for registration of walking and/or running. Rubber tubes connect air chambers in the shoes to the kymograph. The subject wears an accelerometer on his head and holds a bulb for starting the kymograph in his left hand. (From Marey 1874, according to McMahon, T.A. 1984. *Muscles, Reflexes, and Locomotion,* Princeton University Press, Princeton, NJ. Copyright © 1984 by Princeton University Press. Reprinted by permission of Princeton University Press.)

FIGURE 1.3   Ink-spraying outlets according to Vierordt from the 1880s for registering vertical movement components of human locomotion. It is affixed to body segments. (From Vierordt
1881, according to Woltring, H.J. 1977. Measurement and Control of Human Movement. Ph.D.
thesis, Nijmegen: Katholiche Universiteit. With permission.)

locomotion. Vierordt was very interested in clinical work and compared recordings
of healthy subjects and patients with various pathologies, e.g., amputations, tabes
dorsalis, spasticities, and hemiplegies.

As mentioned earlier, Marey first began work on locomotion by applying the
kymographic (odographic) method of registration in 1873. He did this by imple-
menting Carlet's pneumatic method for the registration of foot-floor contacts. As
soon as a sufficiently fast photographic material became available, he moved to photography. It is to Marey that we owe the development of chronophotography. In this
method, a fixed photosensitive plate is intermittently exposed (controlled by a clock
mechanism) and moving objects are shown through successive positions. Using a
rotating disk with one or more apertures in front of the camera lens, one photograph
after another was intermittently exposed, registering a subject dressed in white
against a black background. In this way, it was possible to register the successive
phases of gait, running, jumping, etc. In 1882, Marey published the first chronophotographic pictures of a man marching and a horse jumping over a fence. By using
reflective strips attached to a subject's dark suit, a more abstract representation of
movement trajectories could be obtained. Chronophotography brought great potential to movement analysis since it enabled the direction and speed of change of the
position of particular body landmarks to be determined. Marey's pupil, Demeny, also
helped in development of the chronophotographic method.

As mentioned earlier, at the same time, Muybridge also began his research on
horse locomotion in the U.S. First, the animals themselves triggered cameras by
using wires placed on the ground. At Marey's request, Muybridge tried to study the
flight of birds and in 1881, he brought many photographs to Paris. According to
Marey, however, the aspects of time definition in these observations were unsatisfac-
tory because birds, unlike horses, could not provide triggering themselves. Therefore,
Marey improved Muybridge's method in two ways. First, he constructed a photographic gun. The working principle consisted of the following: a photosensitive plate

consisting of 12 parts of photographic emulsion, when triggered, rotated with high velocity with intermittent stops, giving a series of exposures in 1 s with 1/720 s aperture time. This technical solution is considered to be the predecessor of cinematography. The Parisian astronomer, Pierre Jules César Janssen (1829-1907), used the photographic gun even earlier to record the movement of the planet Venus at a speed of 70/s. With this apparatus Marey could shoot 12 pictures per second, with 1/720 s exposure each, and there was no need for the animal itself to provide triggering. Efforts to increase the number of equidistant exposures (i.e., the frequency of exposures) led to mechanical problems, e.g., pictures were taken on a glass plate of very high mass. Its high inertia limited the number of exposures to 12 per second. The problem was overcome by replacing the glass disk with continuous film thinly covered in gel and silver bromide. The film could be automatically moved in front of the focus of the lens, stopped at each exposure period, and again advanced with a jerk. This resulted in a series of 9 cm^2 photographs of satisfactory dimensions. Marey[21] described this kind of mechanism in detail, and it may be considered, *de facto,* to be a movie camera. By analogy, W. K. L. Dickson constructed a kinetoscope in Edison's laboratory in 1892. The discovery of cinematography by the Lumière brothers followed 3 years later. By using stereophotography, Marey even managed to provide three-dimensional (3D) observations of movement. Savart introduced the procedure of stroboscopy to photographic recording, whereby it became possible to have multiple exposures on the same negative.[16]

In order to have enough light for the photographs, Marey was forced to take his measurements in strong daylight. His successors replaced reflective strips with bulbs or light tubes. This change in technique was named cyclography and was developed to a high level by the Moscow School of Biometrics.

Marey was the first to combine kinematic and kinetic locomotor measurement information. For dynamographic measurement, he used a platform made of oak with so-called "built-in spiral dynamometers" which was based on the principle of elastic deformation of rubber. He combined this platform with an apparatus for measuring the height of vertical body displacement which was connected to the subject's head. Furthermore, he was the first to combine dynamographic and kinematic measurement signals, obtained via photography and presented in the form of a stick diagram.

Two Germans, anatomist Wilhelm Braune (1830-1892) and physicist Otto Fischer (1861-1917), took a decisive step in the study of locomotion in Leipzig in 1895. They perceived the human body as being rigid in the form of a series of dynamic links. They then applied photography and initiated stereometry. In their work, they combined experiments on cadavers, with the goal of determining the inertial properties of body segments, with photographic kinematic measurements of soldiers. Their famous work is "Der Gang des Menschen" (1895) (translated into English by Maquet and Furlong, *The Human Gait,* Springer Verlag, Berlin, 1987). This was the foundation of serious quantitative scientific research of locomotion using photography in 3D space. The relative dimensions of the human body determined by these two authors have been used as a standard to the present time. To increase light intensity during recording, they used special Geissler light bulbs controlled by Rhumkorff coils. They used four cameras: two positioned laterally with respect to the subject and two positioned obliquely frontally. They were the first to describe an accurate, analytical reconstruction of 3D trajectories of characteristic body landmarks from central projections, with an accuracy of several millimeters (!) and 26 frames per second. Their measurement procedure alone, however, lasted 6 to 8 hours, and calculations took several months. This marked the beginning of analytical photogrammetry. Compared to this, Vierordt's method of ink spraying, in spite of great error, was, at that time, more practical for clinical application. Because Braune and Fischer pursued a complete, so-called inverse dynamic approach to the study of locomotion and studied human gait and the earlier-mentioned model of a free double inverted pendulum of the swinging leg, proposed by the Weber brothers 70 years earlier, their work bears great methodological significance. Based on kinematic and kinetic analysis, they showed that during gait on a flat surface the swinging leg was actively controlled by musculature, and was not influenced only by the forces of gravity and inertia as the Weber brothers presumed.[31]

Furnée[16] mentions F. B. and L. M. Gilbreth, who were involved in measuring and studying locomotion from an ergonomic aspect since 1911. Together with F. W. Taylor's principles of organization (1911), work in this field was exceptionally important for the second industrial revolution.

The work of Nikolaj A. Bernstein (1896-1966) in Moscow was also of great importance. He developed and applied precise procedures for measuring human kinematics by using cyclography, a film camera, and a mirror. By slowly moving the film through the camera in the case of repetitive movements, what is known as kymocyclography was obtained. For his 3D analysis, Bernstein abandoned standard stereoscopic cameras because of their low depth accuracy (camera base, 6.5 cm) and used mirror kymocyclography instead. In this method, a mirror was positioned at an angle of 45° with respect to the optical axis of one camera. As a result, each picture included frontal and lateral views of the body whose movement was being studied. This solution circumvented the problem which arose when one needed to time-synchronize the work of two cameras for measurement in 3D space. By applying this method when researching repetitive movement stereotypes, Bernstein reported on errors of spatial position reconstruction smaller than 1 mm per coordinate axis. He increased the sampling frequency (number of pictures) from 26 to 50 to 150 per second. The accuracy in coordinate digitization was 500 μm, which was worse than that of Braune and Fischer (10 μm). He applied a numerical procedure for the mathematical derivation of kinematic data, in contrast to Braune and Fischer who used a graphic procedure. Like Braune and Fischer, Bernstein tried to estimate resultant transferred forces in the centers of mass of the segments. Based on measuring a great number of subjects, he was in a position to test original theories of human motorics and the hierarchical organization of the nervous system, topical even today. His research in the study of movement is epochal since he combined a subtle experimental methodology with the study of the neurophysiological basis of movement, introducing cybernetic concepts. Bernstein developed a hierarchical multilevel model of organization of the system controlling voluntary movement, a theory that had implications later, e.g., in a publication from 1984 (*Human Motor Actions—Bernstein Reassessed,* H. T. A. Whiting, Ed., Elsevier, 1984). His collaborators during the period from 1922 until 1950, Popova, Mogilyanskjayeva, Spielberg, and Sorokin, should also be mentioned.

Rudolph Laban (1879-1958) developed an original method of representing complex human movements. The method, not physical but symbolic (although certain physical terms are used in a descriptive manner), is currently used in dance choreography. This is an approach to represent movement structures complementary to the biomechanical approach.

By introducing a force platform (initiated by Jules Amar in 1916 and later introduced by Eberhard in 1947) and other instruments, it was possible to study human movement more objectively. The predecessor to this device was Marey's dynamographic platform. Elftman (1938) used a plate which could be moved vertically and was suspended on four coil springs, with optical recording of resultant vertical ground reaction force and trajectory of the center of force. Amar introduced the device to measure gait ("The Human Motor," 1920), and Fenn used it to study sprinting (1930, according to Reference 25), while Elftman also measured gait.

At the beginning of the 20th century, electromyography as a technique also began to be applied in the field of locomotion measurement. This was facilitated by the invention of the wire galvanometer. (According to some sources, the term used for this method also stems from Marey.) The invention of the cathode ray tube oscilloscope contributed to further development of electromyography and neurophysiology, and to the understanding of neuromuscular functions.

R. Plato Schwartz of the University of Rochester, Minnesota, is, according to Brand,[3] probably the first medical clinician who developed locomotion measurement methods for clinical, and not primarily research, applications in the 1930s and systematically collected and analyzed gait measurement data. He also postulated the basic requirements for the gait measurement method. (Those presented in Chapter 3, Section 3.3, are to a great extent similar to his.) Schwartz measured a great number of patients using a device similar to Carlet's, but his device consisted of three—not two—air chambers. He applied the electrobasograph, an instrument equipped with three electrical contact switches, on the lower surface of the heel and the first and fifth head of the metatarsal.

In the middle of the 1940s, the Berkeley Group led by Saunders, Inmann, and Sutherland began their work in the Biomechanics Laboratory at the University of California at Berkeley. Their field was orthopedic rehabilitation of individuals who had been injured in the Second World War and who primarily needed prostheses for their extremities. They developed various movement measurement techniques: kinematic, kinetic, and electromyographic. In kinematic measurements via photography, classical stereometry was not used, as first defined by Braune and Fischer and later by Bernstein, but a more direct, practical method, which will be presented in the description of the photographic kinematic measurement method in Chapter 4.

Stroboscopic photography with multiple exposures and cinematography were dominant techniques to measure human kinematics until the 1970s. The perfection of stroboscopic photography as a technique may be illustrated by an example from the collection of Harold E. Edgerton (1903-1990), an eminent expert in this technique at MIT in Cambridge, Massachusetts (Figure 1.4).

Further development of locomotion measurement systems was characterized by

FIGURE 1.4 Stroboscopic photography of a tennis shot from the beginning of the 1970s, taken at a speed of 25 frames per second. (Taken by Harold E. Edgerton.)

an ever greater influence of technology and engineering. In the 1970s, through the introduction of digital computers, measurement procedures were automatized to a significant degree, becoming more efficient. Development of the fields of semiconductor physics, electronics, measurement techniques, automatic control, telemetry, video and consumer electronics, and computing and computer graphics continuously contributed to new solutions of measurement, quantitative evaluation, and diagnostics of locomotion. Development in this field was also marked by the formation of international professional societies, the most important of which was the International Society for Biomechanics, and the publication of professional periodicals, such as the *Journal of Biomechanics,* the *Journal of Biomechanical Engineering, Human Movement Science,* etc. The first international biomechanical conference was held in 1967 in Zürich, Switzerland (initiator E. Jokl; organizer J. Wartenweiler), and the first world congress in biomechanics was held in 1990 in San Diego, California, (chairman Y.C. Fung).

Most of the information in this historical survey has been devoted to methods for measurement of locomotion kinematics. Since this type of measuring closely resembles a human's natural visual perception of moving objects, this aspect of motion study has, understandably, been of greatest interest to research. Reviews on the historical development of some kinetic methods will be presented in Chapter 5; myoelectric measurement methods will be presented in Chapter 6.

2 Methodological Background

This chapter covers basic methodological issues in the study of human locomotion, in which the measurement of kinematic and kinetic (Section 2.1) and myoelectric (Sections 2.2 and 2.3) quantities are relevant. These methodological issues offer a theoretical framework in which the results of measurement methods—to be described later—may be interpreted. Biomechanical aspects are presented in the form of a comprehensive summary in Sections 2.1 (the body as a whole) and 2.3 (the skeletal muscle). Neurophysiological aspects of locomotion are also described, illustratively (Section 2.2).

2.1 BIOMECHANICAL MODELING OF THE HUMAN BODY AND THE INVERSE DYNAMIC APPROACH

Implicit to locomotion measurement is an appropriately simplified and idealized representation, i.e., modeling of the body. The purpose of modeling is to isolate crucial mechanical features of the body from the standpoint of performing movements, while at the same time ignoring everything which in the first approximation is unimportant for locomotor activities. Since locomotion occurs in space and time, the moving body, like any other material object, is subject to influences governed by the laws of classical mechanics. When studying the movement of the body as a whole—the primary objective of this book—the appropriate idealization is a mechanical body model composed of rigid segments interconnected by joints. "Inertial parameter" models of this kind have been used in a number of studies on locomotor activities.

The fundamental principles of the inverse dynamic approach, which calculates kinetic (dynamic) movement quantities, forces and moments of force, based on known kinematic data, acquired through measurement, will be presented. Mathematical assessments are according to References 33 through 35.

Based on the measured anthropometric (morphologic) body dimensions of a particular subject, inertial features of his body segments (the upper leg, the lower leg, the trunk, etc.) may be estimated. In doing so, two approaches are traditionally used. The first approach applies regression equations which relate measured dimensions with estimated parameters, obtained by statistical processing of data measured on human cadavers—the regression approach. In the second approach, inertial parameters are applied, and it is based on simplifications, i.e., geometrical and material idealizations of particular body segments. (There is also a third approach, whereby direct measurements are applied and obtained through computer tomography, CT, or magnetic resonance imaging, MRI.[36,37] However, these methods are expensive and, in addition, some of them are also undesirable due to radiation (CT). The 3D laser scanner is yet another possibility.) Application techniques, accuracy, and other features of the corresponding methods are not treated here. Further details may be obtained in classical works on biomechanics.

As a result of these estimates, the following inertial parameters are assigned to each body segment, rigid and of homogenous density:

- Mass (m)
- Position vector of the center of mass (CM) in the coordinate system (x,y,z), fixed to the segment (This coordinate system is called local, while the position vector is marked $\mathbf{p}_g$.)
- Principal axes of inertia defined relative to the local coordinate system through columns of the transformation (rotation) matrix [B]
- Moments of inertia around the principal axes passing through the mass center (I_x, I_y, I_z)

Determination of these quantities presupposes that the local coordinate system of the segment has been defined.

Since each body segment moves through space during locomotion, the absolute position of any body point in the global coordinate system (X, Y, Z), fixed in relation to the observer, may be expressed by the following equation:

$$\mathbf{P} = [A]\,\mathbf{p} + \mathbf{P}_o \qquad\qquad (2.1)$$

where $\mathbf{p}$ denotes the position vector of any point in the local coordinate system

$\mathbf{P}$ denotes the position vector of the same point in the global coordinate system

[A] denotes the transformation (rotation) matrix

$\mathbf{P}_o$ denotes the position vector of the origin of the local coordinate system with respect to the global coordinate system

The columns of the matrix [A] are the direction cosines of the local with respect to the global coordinate system.

The body is represented by multiple connected rigid segments which, analogous with the conventions of robotic mechanisms, form a kinematic chain. Problems concerning the geometrical and mechanical descriptions of particular joint connections within this system will not be treated here, but will be commented on later. In general, however, the resultant mechanical interaction between any two neighboring segments may be expressed by a three-component force vector and a corresponding moment vector. This is also true of interactions between bordering segments (typically the foot) and their surroundings.

A general free-body diagram of the n-segment mechanical system is depicted in Figure 2.1. Corresponding force and moment vectors acting at particular points of contact are marked. Points R_s (s = 1,n) denote positions where, it is supposed, total resultant forces and moments of forces between segments are reflected. Point O is an arbitrary point to which the intersegmental forces have been reduced.

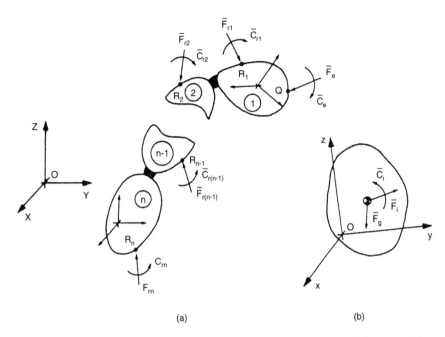

FIGURE 2.1   (a) A general free-body diagram of the n-segment mechanical system; (b) one segment. (From Cappozzo, A., *Biomechanics of Normal and Pathological Human Articulating Joints,* N. Berme, A.E. Engin, and K.M. Correia de Silva, Eds., Kluwer, Dordrecht, The Netherlands. With kind permission from Kluwer Academic Publishers.)

The dynamic equilibrium of the free body is described by the following system of equations:

$$\mathbf{F}_g + \mathbf{F}_i + \mathbf{F}_r + \mathbf{F}_e = 0 \tag{2.2}$$

$$\mathbf{M}_g + \mathbf{M}_i + \mathbf{M}_r + \mathbf{C}_i + \mathbf{C}_r + \mathbf{C}_e = 0 \tag{2.3}$$

Symbols denote:
    $\mathbf{F}_g$ = resultant vector of gravitational force
    $\mathbf{F}_i$ = resultant vector of inertial force
    $\mathbf{F}_r$ = resultant vector of external reaction force
    $\mathbf{F}_e$ = resultant intersegmental force vector
    $\mathbf{C}_i$ = resultant vector of inertial reaction couple (couple = moment of force)
    $\mathbf{C}_r$ = resultant vector of external reaction couple
    $\mathbf{C}_e$ = resultant intersegmental couple vector
which influence n segments.

    $\mathbf{M}_g$ = resultant moment vector of gravitational force
    $\mathbf{M}_i$ = resultant moment vector of inertial force
    $\mathbf{M}_r$ = resultant moment vector of external reaction force
which influence n segments and are calculated with respect to point Q.

Gravitational and inertial force vectors which influence particular segments in relation to the global coordinate system are given as follows (the subscript designating the segment has been omitted):

$$\mathbf{F}_g = m \begin{bmatrix} 0 \\ 0 \\ -g \end{bmatrix} \tag{2.4}$$

$$\mathbf{F}_i = m\ddot{\mathbf{P}}_g \tag{2.5}$$

with $\mathbf{P}_g = [A]\,\mathbf{p}_g + \mathbf{P}_o$

The inertial couple vector in relation to the global coordinate system is given as follows:

$$\mathbf{C}_i = -[A][B] \begin{bmatrix} I_x\,\dot{\omega}_x + (I_z - I_y)\,\omega_y\,\omega_z \\ I_y\,\dot{\omega}_y + (I_x - I_z)\,\omega_x\,\omega_z \\ I_z\,\dot{\omega}_z + (I_y - I_x)\,\omega_x\,\omega_y \end{bmatrix} \tag{2.6}$$

where ω_x, ω_y, and ω_z are the segment angular velocity vector components relative to the principal axes of inertia. They may be expressed as a function of the angular velocity components in the local coordinate system as follows:

$$\begin{bmatrix} \omega_x \\ \omega_y \\ \omega_z \end{bmatrix} = [B]^T \begin{bmatrix} \omega_{xl} \\ \omega_{yl} \\ \omega_{zl} \end{bmatrix} \tag{2.7}$$

where:

$$\omega_{xl} = \dot{a}_{12}\,a_{13} + \dot{a}_{22}\,a_{23} + \dot{a}_{32}\,a_{33} \tag{2.8}$$

$$\omega_{yl} = \dot{a}_{13}\,a_{11} + \dot{a}_{23}\,a_{21} + \dot{a}_{33}\,a_{31} \tag{2.9}$$

$$\omega_{zl} = \dot{a}_{11}\,a_{12} + \dot{a}_{21}\,a_{22} + \dot{a}_{31}\,a_{32} \tag{2.10}$$

and where a_{ij} and $\dot{a}_{ij}$ are elements of the matrix $[A]$, i.e., their time derivative, respectively.

To calculate $\mathbf{M}_g$, $\mathbf{M}_i$, and $\mathbf{M}_r$, the position vectors of points Q and R_s ($s = 1, \ldots, n$) in the global coordinate system are needed. The former vector is usually given in the local coordinate system and so has to be transformed to the global coordinate system by using Equation 2.1. The second vector is usually given directly in the global coordinate system.

So, if the time functions of the positional variables ($\mathbf{P}_o$, $[A]$), the inertial parameters (m, $\mathbf{p}_g$, $[B]$, I_x, I_y, I_z), and the external reaction vectors $\mathbf{P}_r$, $\mathbf{F}_r$, and $\mathbf{C}_r$ are known for each model segment, and if the local position vector $\mathbf{p}_g$ is known as well, then the intersegmental force and couple vectors ($\mathbf{F}_e$ and $\mathbf{C}_e$) can be calculated using the earlier mentioned equations. This is what is known as the inverse dynamic approach. The aspects of its practical realization shall be discussed within the context of specific measurement methods in Chapters 4 and 7.

Furthermore, each segment's and the entire body's potential and kinetic energy may be calculated. In this way, the total organism's energy expenditure caused by locomotion may be characterized from the mechanical standpoint.

In order to determine the 3D spatial position of a particular observed point experimentally, a stereometric procedure has to be performed. The corresponding procedure is subject to kinematic measurements (described in Chapter 4.2). With the help of 3D positions of three known noncolinear points on the rigid body, the spatial position of this body is completely determined. In applying this principle, and bearing kinematic measurements in mind, markers are attached to the subject's body. It is customary to position either single markers or three-marker clusters fixed to a rigid support, i.e., arranged in a known mutual geometrical relationship. Alternatively, if no markers are used, positions of characteristic body landmarks have to be extracted later from recordings obtained. When carrying out a specific measurement method, i.e., procedure, we have to be familiar with geometrical relationships between selected marker locations and anatomical body landmarks, such as imaginary centers of rotation of particular joints, etc. One practical possibility is illustrated, using three-marker clusters fixed to a rigid support (Figure 2.2).[38]

Problems concerning the geometrical and mechanical description of particular joints of the human body, as mentioned above, may be addressed in many ways and with various degrees of simplification. The simplest model is the hinged (ball-and-socket) joint which might, for instance, be adequate for the knee if only one-axial rotation is assumed; more complex models are sliding, rolling, and spinning joints. The situation is complex from the anatomical point of view. The so-called screw axis of motion may be considered a method of choice, which, at the level of global human body kinematics, is the most realistic anatomically and which is described at any point in time by six parameters that may be calculated by using time functions $\mathbf{P}_0$ and [A] connected with any two segments. (Woltring calls this instantaneous helical axis or instantaneous axis of rotation.) At each point in time, movement of the joint is visualized as the movement of one body segment with respect to a neighboring segment, where movement is considered to occur as a translation along a spatially directed line. The position and direction of this line may vary in time.

The above described concept is the general 3D form of the inverse dynamic approach, without taking any specific kinematic measurement method into consideration. In practice, one may pursue special cases of the approach in two dimensions (2D), i.e., planar, which are sometimes adequate approximations for planar movements (long jump). At this point, problems of numerical calculation of the mentioned quantities will not be addressed. In fact, since Braune and Fischer, it was not until the rapid development—and consequent decrease in price—and widespread application of digital computers that the practical application of physical approximations of body movements in the study of locomotion was made possible. The systematic construction of mathematical models of bipedal locomotion and corresponding stability criteria in the sense of gait synthesis took place in the 1970s in the works of Stepanjenko and Vukobratović.[39,40]

FIGURE 2.2 One possibility for positioning body markers in (clinical) measurements of locomotion kinematics. (From Davis, R.B., De Luca, P.A., and Õunpuu, S. 1995. *The Biomedical Engineering Handbook*, Bronzino, J.D., Ed., Boca Raton, FL: CRC Press, 381–390. With permission.)

The complexity of the chosen biomechanical model is determined by the application intended. For instance, in studying nearly static body postures, models composed of only a few segments often suffice. So, for example, in the study of ground reaction force resulting from mechanical body oscillations during aiming in sport rifle shooting, i.e., a quasi-static activity, a simple one-segment body model, the so-called inverted pendulum used by Nashner and Gurfinkel (according to Reference 41), was suitable. The study of natural locomotion, however, usually requires models consisting of seven to eight segments at least, the lower extremities being represented quite faithfully (the upper leg, the lower leg, the foot), while the total upper body torso may be represented by just one segment. Calculation of the dynamics of even simple mechanical systems is considerably demanding.

The biomechanical model of the body of an inertial type according to Hanavan[42] had 15 geometrically shaped segments. This model, first developed for space research, i.e., ergonomic man-vehicle studies, is suitable for a number of locomotion studies.

Researching biomechanical characteristics of particular parts of the locomotor system in more detail requires a more precise modeling of anatomical properties. A few complex efforts to faithfully model the configuration of specific joints are worth mentioning. In one approach, it was possible to determine the spatial distribution of forces at knee joint surfaces.[43] Kaufmann and An[44] provided a systematic and concise review of mechanical models of joint surfaces in all major body joints: ankle, knee, hip, shoulder, elbow, wrist, and hand. They used the earlier mentioned concepts of kinematic modeling of joints (rolling, sliding, etc.), combining them with known anatomical facts. Each joint's specific characteristics determine its respective musculoskeletal function. A knowledge of joint characteristics with regard to the range of movement, stability, etc. determines, in practice, the joint's susceptibility to injuries, degenerative changes, etc.

Hatze developed a complex biomechanical model of the whole body (Figure 2.3).[39,45] It consists of 17 segments, anthropomorphically designed and linked by joint connections of various complexity. 3D movements are possible and the total number of mechanical degrees of freedom of such a system is 44. Furthermore, modifications in some model parameters according to sex are allowed. Since there are 242 input anthropometric measures required, the application of the model is very time consuming. An anthropomorphic, humanoid form of the model is further enhanced by the actuator elements, the muscles, as well as by the control elements, determined by neural signal features controlling muscular activity, making the model not only biomechanical, but myocybernetic as well. (Hatze, in fact, postulates myocybernetics as an area of application of cybernetic principles to the neuromuscular system.) By applying the nonlinear differential equations system, and relying on experimentally determined values of particular physiological parameters of the skeletal muscle model from his own research and the research of others, the author has described the complete movement dynamics of such a system and developed corresponding software in FORTRAN. The model aims at simulations of dynamic behavior and studies of optimization principles governing the function of the human neuro-musculo-skeletal system. Hatze's model, however, in spite of its comprehensiveness, but because of its high degree of complexity and high price, did not find wide application at the time.

On the level of global body modeling, a more precise representation of anatomy may be seen in the approach by Delp et al.[46] A faithful geometrical model of the lower extremity skeleton was stored in a computer's memory and the muscles, 43 in all and represented by dominant lines of action, were superimposed on it. With the help of computer graphics, simulations of different system positions with calculation of supposed muscular forces in isometric conditions, as well as the simulated interventions in the system (such as tendon transfer), were made possible, making the model an original and valuable learning vehicle for orthopedic surgeons. By using models of this kind, it is possible to provide estimation of forces of particular muscular/tendon structures participating in the formation of the total force. In this way, the indeterminacy in defining the function of neuro-musculo-skeletal system is reduced, a function which was previously solved by applying mathematical optimization methods. However, as has been said, the values resulting from the inverse dynamic approach

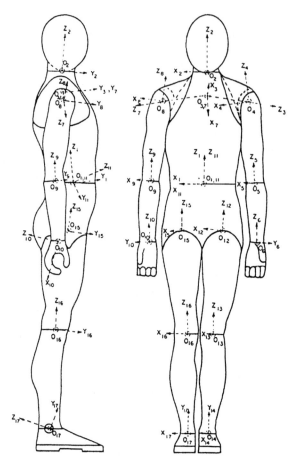

FIGURE 2.3 Biomechanical inertial model of the human body: rigid segment type. (From Hatze, H. 1980. *IEEE Trans. Automatic Control.* 25:375–385. © 1980 IEEE. With permission.)

($\mathbf{F}_e$ and $\mathbf{C}_e$, Figure 2.1) are resultant values, while particular components participating in the formation of resultant forces and moments may attain significantly larger absolute values, which is of potential interest, affording insight into traumatic effects.

Due to the complexity of the biological system being modeled, the development and application of biomechanical models of the human body are important areas in the kinematics and dynamics of multisegmental mechanical systems in general. While models like Hanavan's are certainly very simplistic, due to the decrease in the price of computers and the development of methods of noninvasive visualization of geometrical properties of biological tissue, CT and MRI, significantly more complex and faithful approaches can be applied in practice today.

Disregarding the level of complexity, what is common to all biomechanical body models is that they represent a necessary means for the application of the inverse dynamic approach. Kinematic variables describing the movement geometry of such a system through defined landmark body points is a starting point. Kinetic (dynamic) variables are obtained by the inverse dynamic approach. Kinematic variables themselves only represent the description of movement and do not possess great informational content. Some of the external kinetic data, such as ground reaction force and moment of force ($\mathbf{F}_r$, $\mathbf{C}_r$), may be directly measurable as well. If the movement of the body as a whole is being analyzed, such measurement data have to be combined with those mathematically estimated and, hence, may be used in the evaluation of the inverse dynamic approach (Chapters 4 and 5).

2.2 NEUROPHYSIOLOGY OF LOCOMOTION

The anatomical substrate of the locomotor system is made up of bones with their surrounding tissues, such as cartilage, muscles, ligaments, the nervous system controlling the motorics, and connective tissue. The skeletal subsystem, supporting the body as a whole, may be called the effector subsystem. The skeletal muscle plant, active force generators with the purpose of realizing movement, may be called the actuator subsystem. Active muscular forces combined with external forces (gravitational, ground reaction), elastic muscle forces, and other inertial forces which arise due to the moving body mass, all determine manifested body kinematics in time and space. The subsystem, called control, which coordinates and controls overall motor activity, is the nervous system. The names of the subsystems used characterize the technical, bioengineering approach to the study of the locomotor function.[39]

The functional-anatomical substrate of the neuro-musculo-skeletal system is described in pertinent literature.[47,48] In the bioengineering approach, the goal is to reach an idealized and highly reduced representation of the anatomical structure of the locomotor apparatus. For example, the lower extremity musculoskeletal structure may be presented schematically as reduced to ten dominant muscular groups responsible for flexion and extension of lower extremities in the sagittal plane (Figure 2.4). According to the muscle equivalent concept,[49] each synergistic muscle group is substituted by one muscle, representative of its function. Various degrees of approximation of the real functional-anatomical structure of the neuro-musculo-skeletal system are possible, which have been adapted to the study of a specific locomotor activity. On the other hand, the degree of complexity and fidelity of the model are often a practical result of technical possibilities for measuring and data processing available. In view of the rapid development of computers, the tendency is to apply anatomically correct models in three spatial dimensions; the model by Delp et al.[46] is an example of this approach.

As active force generators in the realization of movement, skeletal muscles are coordinated in time and space and through contraction intensities when a particular movement structure is realized. According to Hess (1954, according to Reference 50): "The course of a movement is nothing else but a projection to the outside of a pattern of excitation taking place in a corresponding setting in the central nervous system." The nervous system, acting on the musculoskeletal, is an adaptive structure

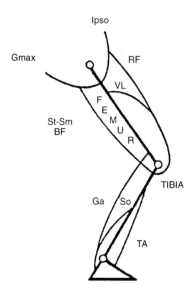

FIGURE 2.4 Lower extremity musculoskeletal system's anatomy highly simplified. The symbols denote: Ipso, musculus iliopsoas; Gmax, m. gluteus maximus; RF, m. rectus femoris; VL, m. vastus lateralis; St-Sm, m. semitendinosus - m. semimembranosus; BF, m. biceps femoris; Ga, m. gastrocnemius; So, m. soleus; TA, m. tibialis anterior.

that has acquired certain activity patterns during phylogenesis which develop during growth. In this way, stereotypical (loco)motor activities, such as walking, manifest, to a great extent, the automatism of excitation and inhibition patterns and, in concordance with this, the defined order and contraction intensity of the corresponding musculature. Contrary to this, new types of locomotion unusual for humans and even an athlete in good condition, such as acrobatic elements in gymnastics, do not initially manifest automatism of muscular activity patterns, but are acquired gradually through multiple repetitions, exercise, and training. A decrease in and rationalization of energy expenditure necessary for metabolic processes in the muscles are a part of the process.

In disturbed function of the locomotor apparatus due to disease or injury, the natural automatism of stereotypical locomotor activities is disrupted. Rehabilitation then aims at restoring the disturbed function or, in the case of permanent anatomical or functional changes or deficiencies, modifying the function of the neuro-musculo-skeletal system so that the role of the insufficient or lost muscles is taken over by the remaining (healthy) musculature. Hypothetically, the motor coordination stereotype has to be restructured. In healthy and pathological locomotion, a motor learning process takes place and a proprioceptive memory is formed.[51] To illustrate, some basic functional mechanisms of the neuromuscular system fundamental to the realization of natural locomotion are pointed out.[22,52]

2.2.1   RECIPROCAL INHIBITION

Reciprocal inhibition means the stretching of the extensor muscle in a particular joint inhibits the activity of motor neurons innervating the flexor muscles in the same joint and *vice versa*. This is accomplished through the interneuron, located in the spinal cord, which "reverses the sign" of the incoming afferent message from the muscle first mentioned, thus inhibiting the second one. In this way, the simultaneous contraction of antagonists in a joint in the presence of external forces is prevented and the functioning of the muscular circuit is rationalized. Examples of this action are musculus biceps brachii and musculus triceps brachii of the upper arm acting in the elbow joint. Similar mechanisms also exist in other body joints. For example, in the lower extremities, this effect is displayed in walking, running, and similar locomotion during ground reaction force.

2.2.2   THE PLACING REACTION AND REFLEX REVERSAL

The placing reaction is an automatic movement, "wired" at the level of the spinal cord, which occurs when the upper (dorsal) part of the foot is touched or stimulated causing flexion of the leg if it is in the air (swing phase), i.e., extension of the leg if it is in the supporting position. Reflex reversal occurs slightly prior to leg-floor contact, so that the "trigger" certainly is not ground reaction force, but the sensor organs in muscles and joints.

2.2.3   AUTOMATIC GENERATION OF LOCOMOTOR PATTERNS

This feature concerns the function of the sensomotor neurons in animals that control the cyclic movements of extremities independent of higher nervous centers. So, for example, decerebrated (spinal) animals reflexively maintain a locomotor stereotype even when the connection with higher nervous centers is completely broken. It has been suggested that the spinal cord contains groups of neurons which are arranged in such a way that they reciprocally activate or inhibit each other in a rhythmic fashion.[7] Grillner (1985, according to Reference 7) indicates the existence of "central pattern generators," i.e., neurons controlling specific activities via groups of other neurons, organized hierarchically up to the spinal cord level (the control pyramid). Such neuron pools are called oscillators and are responsible for the timing of many movements including locomotion. In humans, this mechanism is lost during ontogeny.[53] Comparative locomotion studies in animals in general have indicated some distinguishing traits in humans, such as the initial impulsive heel impact and the "plantigrade" gait, i.e., a mode of locomotion where, after initial heel impact, the whole foot is placed on the floor (Gray 1968, according to Reference 25).[54] Animals largely manifest a different pattern of stance phase during locomotion by establishing either full contact during initial contact or by establishing initial contact with the front part of the foot. In children, as well as in individuals with Parkinson's disease, this mechanism recurs.[25]

2.2.4   HIERARCHICAL ORGANIZATION OF MOTOR CONTROL

The central nervous system's (CNS) control of motorics and locomotion is exceptionally complex. A significant contribution to this field came from Charles Scott Sherrington (1857-1952) in Great Britain. He introduced the terms proprioception and synapse and pointed to a mechanism by which complex motor acts may be produced by connecting a large number of simple reflexes. Today, due to Bernstein's research and research that followed, this field has branched into a rich multi- and interdisciplinary field which includes neurophysiology, neuroscience, several clinical medical areas, cybernetics, computer science, etc. For now, it suffices to state the components in the hierarchical system of motor control: motor cortex, basal ganglia, thalamus and hypothalamus, midbrain, cerebellum, brainstem, and spinal cord.

A fact that needs to be stressed is that sensory and motor functions are connected at all levels which is a feature of CNS. In other words, there are very few sensory programs that do not finish with the initiation or modification of movement. The sensory functions include not only proprioceptive organs in the neuromuscular system (muscle spindles and Golgi tendon organs, Chapter 2.3), but also visual, auditory, and vestibular system functions. Figure 2.5 is a scheme of the sensomotor mechanism of the neuromuscular locomotor system.[55]

As far as motor learning is concerned, an important term is the motor engram. Gage[7] cites Kottke's (1980) definition of the motor engram as a "pathway of

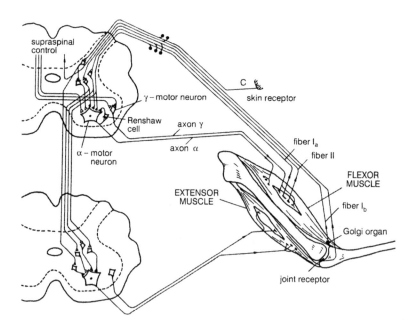

FIGURE 2.5   Scheme of some neuro-musculo-skeletal sensomotor mechanisms; example of the upper arm. (From Vodovnik, L. 1985. *Nevrokibernetika,* Ljubljana, Slovenija: Fakulteta za Elektrotehniko. With permission.)

interneuronal linkages involving activation of certain neurons and muscles to perform a pattern of motor activity in a specific sequence of speed, strength, and motion, and at the same time inhibition of other neuron pathways so that muscles which should not be participating in this pattern remain quiet." If the practiced activity has been precise, the engram will also be precise. Kottke describes the CNS organization as a reflex syndrome with neuronal activation resulting due to a response to external stimuli. The "voluntary system of control" is automatically regulated to maintain the system in a state of excitation which is just below the threshold. Voluntary activation adds just enough impulses to reach excitation level which would lead to motor neuron discharge. The only part of this system over which we have direct control is the limited-channel excitatory pathway which is the pyramidal tract from the motor cortex. Therefore, neuromuscular function is based on reflexes, and cerebral modifications of motor function are integrated through the basal ganglia, cerebellum, and brainstem to produce multimuscular coordination through selective patterns of excitation and inhibition of supraspinal and spinal reflexes.[7]

Therefore taking the pathology of the nervous control of motorics into account, damage to the highest or intermediate centers will cause abnormalities in performance by releasing the activity of the undamaged center at the next level down from control, rather than by generating a new form of activity originating from the damaged center. Kottke further states that an infant, it seems, captures control of reflexes by first learning to facilitate reflex activity via the pyramidal excitatory mechanism. At first, the activity is crude, random, and essentially ineffective. As the association between voluntary excitation and reflex excitation is repeated, the infant becomes aware of the response. Voluntary excitation becomes increasingly effective until eventually the child can produce a desired muscular activity without manifest reflex participation. The maturating of human locomotor pattern can be considered to be a process by which spinal circuits are progressively increasingly influenced by higher brain centers (Forsberg 1985 and Dietz 1987, according to Reference 53). Ulrich et al. (1991, according to Reference 7) quite rightly state, "This ontogenetic scenario supports a view of adult locomotion also as an adaptable, emergent coordination rather than one driven primarily by centrally determined pattern generators—independent locomotion emerges in a self-organizing manner from a system predisposed to generate an alternating pattern, but whose details are not specified. These details are carved out through interactions with the periphery and in particular, from the biodynamic field."

Elaborating the problem of learning, i.e., acquisition of movement stereotypes in sports, Rieder[56] states that to acquire a certain movement structure, a javelin thrower, for example, needs to throw a javelin about 50 times to get the general feel of it, 500 times to acquire the movement, and 5000 times to perfect the movement.

2.2.5 COMPUTATIONAL NEUROSCIENCE AND LOCOMOTION

In view of what has been stated, it is evident that motor control of locomotion is exceptionally complex and so, at a certain level, a more abstract approach has to be pursued. One possibility is offered by the field of computational neuroscience in which the nervous system is theoretically studied by applying a "top-down" approach which attempts to discover which computational operations are really needed for the realization of its function. Within a broader context, such an approach belongs to artificial intelligence. Of course, the applicability of computer simulations is of great practical importance in this field. This reflects the connection with engineering areas, such as theory of automatic control, automatics, robotics, etc.

In motor control research, however, a direct analogy between neural and computer processing cannot be established. As opposed to the complementary area of vision research, as yet there is no consensus on the fundamental transformations needed. It is not known, for example, whether control influences variables at the level of muscle, joint, or final point of movement or if the variables specify rigidity, length, speed, force, or moment. While, in essence, the feature of vision consists of processing information, a feature of motorics does not simply signify processing of information, but encompasses biomechanical properties of the body, i.e., the organism (defined in subsections 2.1 and 2.3), which are *de facto* interpolated between nervous control and motor (behavioral) performance. Limitations at a neuromuscular, mechanical, as well as an external level can be defined. Furthermore, research may take place at the level of object, joint, or actuator. An approach to these issues is to study these complex systems within the framework of the fields of mathematics, areas such as synergetics, nonlinear systems theory, etc. A connectionist approach is also topical, i.e., via neural networks, characterized dominantly by the feature of adaptivity.

Although Bernstein pointed out many general dilemmas in this field, progress today greatly depends on technical and technological development. A special problem in research is the difficulties in measuring movement, the primary concern of this book. General principles of motor control have not been established.[57] Conceptually, the schema according to Winters[58] (Figure 2.6) may be useful. It shows a simplified version of information processing in the neuro-musculo-skeletal system. It is worth stressing that general principles which would relate to all movements do not exist, but that nervous and skeletal systems are adapted to perform a number of tasks relatively well and not just a few perfectly. All further considerations, however, are focused on the peripheral part of the nervous system—on the skeletal muscle in particular, as it is the principal movement actuator and a source which can be measured through EMG signals.

2.2.6   THE PERIPHERAL NEUROMUSCULAR SYSTEM

The basic functional block of the lowest peripheral part of the neuromuscular system, a skeletal muscle control circuit, might, schematically and didactically simplified, be as repesented in Figure 2.7.[59] One can identify the path of active stimulation of the muscle via the motoneuron. In addition, feedback from receptors sensitive to contraction speed and muscle length (muscle spindles) and receptors sensitive to the strain (Golgi tendon organs) can be identified.[60] Sensitivity of the system of muscle spindles is adjusted with the help of γ neurons. This type of functional block is incorporated within all complex sensomotor systems of movement realization.

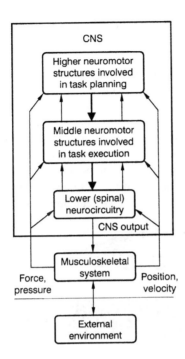

FIGURE 2.6 Simplified representation of information processing in the neuro-musculo-skeletal system. (From Winters, J.M. 1995. *Three-Dimensional Analysis of Human Movement,* P. Allard, I.A.F. Stokes, and J.-P. Blanchi, Eds., Champaign, IL: Human Kinetics, 257–292. With permission.)

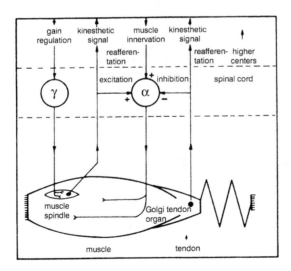

FIGURE 2.7 Skeletal muscle peripheral control circuit. (From Beulke, H. 1978. *Leistungssport* 3:224–235. With permission from Philippka-Verlag.)

2.3 BIOMECHANICAL MODELING OF SKELETAL MUSCLE

A summary of the anatomy of the skeletal muscle and its biomechanical functions is now given. For more details see McMahon.[22]

Skeletal muscles are the active force generators, the so-called actuators within the locomotor system. Muscular tissue is characterized by excitability, conductivity, elasticity, and above all contractility, enabling it to function as actuator. There are hundreds of skeletal muscles in the human body and each of them fulfills a specific function. A muscle consists of a large number of cells, elongated in form: muscle fibers, 10 to 100 μm in diameter and up to 30 cm in length in their relaxed state. Figure 2.8 illustrates the structure of skeletal muscles. Fibers are grouped into

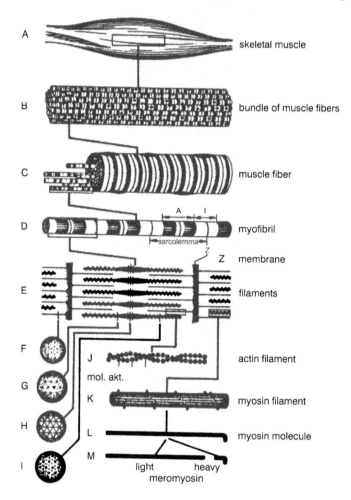

FIGURE 2.8 Skeletal muscle anatomy. (From Vodovnik, L. 1985. *Nevrokibernetika,* Ljubljana, Slovenija: Fakulteta za Elektrotehniko. With permission.)

bundles and several bundles form a muscle. Fibers end in tendons through which a muscle is connected to the bone. Muscular fibers are covered by a sarcolemma membrane, and they consist of tiny threads, myofibrils, protein structures of a specific form. Their basic structure consists of two types of fibers—thin actin and large myosin fibers. Thin fibers are connected in the Z-membrane, and the part of the myofibril which is separated by two membranes is called the sarcomere, the fundamental functional unit of myofibril. The form and function of myofibril enable muscular contraction, through specific electrical, chemical, and mechanical conversion, by sliding filaments, a shortage of sarcomere.

While itself a unity within the neuro-musculo-skeletal system, a unity within one muscle is the motor unit. The term comes from Liddel and Sherrington (1925) and describes a structure consisting of one efferent motoneuron (α) and a group of muscle fibers innervated by this neuron. The number of innervated fibers may vary from a few, as in some muscles active in "fine" movements, to several thousand, as in the lower extremity muscles. The place of innervation, a specifically formed synapse, is called the neuromuscular plate or neuromuscular point.

The motor plate is usually located around the center of a fiber in a longitudinal direction (Figure 2.9). An incoming nerve impulse causes an electric change in the neural terminal leading to the release of a transmitter (acetylcholine) which binds itself to muscle cell receptors, resulting in the depolarization of the sarcolemma. Depolarization occurs in both directions along a muscle fiber. Consequently, an electromagnetic field is generated in its environment. Because the final global effect is our concern, the complex electrochemical mechanism shall not be discussed in further detail. The total bioelectric signal of a muscle is a result of the spatiotemporal

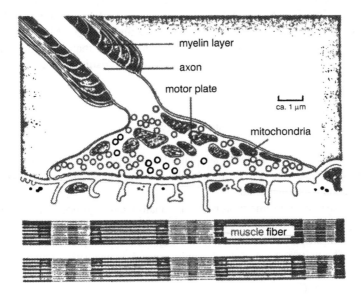

FIGURE 2.9 The motor plate. (From Vodovnik, L. 1985. *Nevrokibernetika*, Ljubljana, Slovenija: Fakulteta za Elektrotehniko. With permission.)

summation of activity of a large number of motor units (interference pattern). Muscle action potentials "accompany" chemical processes where mechanical and thermal energy is released. According to Katz,[61] "Muscular action potential realizes a fast mobilisation of the contractile apparatus in the interior of the muscle cell." The propagation speed of muscle action potentials might be in a 2- to 6-m/s interval, while impulse width is 1 to 5 ms. In Chapter 6, the issue of modeling myoelectric processes and the mathematical processing of surface EMG signals will be addressed in connection with EMG signal measurement.

From the systems' standpoint, the electrophysiological process of stimulation and contraction of skeletal muscles[61] can be taken as being the transformation of information contained in a train of electrical impulses, i.e., pulse-coded information traveling along the neural fiber, into a mechanical force. Apart from the mechanical effect, as mentioned, an electrical change occurs in the muscle's state through the propagation of electrical impulses in the motor unit. The parallelism in these two processes, where an electrical precedes a mechanical, provides a basis for the indices of electrical events in a muscle to serve as indicators of a developed mechanical force. This is the basis for the kinesiological importance of myoelectric signals which are to be found at the muscle surface and are therefore more easily accessible for measurement (Chapter 6).

As carrier signals of control (or sensory) information in the neuromuscular system, neural action potentials are equal in shape, while the transmitted information is contained in time intervals between the successive action potentials, i.e., in engineering terms, pulse-frequency modulation occurs. The propagation speed of neural action potentials may be within a 0.5- to 120-m/s interval.[47]

The electrical processes of propagation of the action potential in a neuron and then in muscle fiber lead to the realization of a muscle twitch with a delay of a few milliseconds (latency time). A twitch is defined as a time change of isometric force. The twitch phenomenon lasts an order of magnitude longer than the electric action potential. Muscle fibers differ in regard to twitch characteristics, so there are:

- Fast muscle fibers (≤ 50 ms)
- Slow muscle fibers (~ 120 ms)
- Muscle fibers with characteristics in-between

(Due to the neuromuscular system's adaptability, the type of neural stimulation may consequently influence the type of muscle fibers. There are examples of experiments of this type conducted on animals.)

Muscle fibers of the same motor unit are anatomically grouped into subgroups that are dispersed throughout the muscle. Therefore, motor neuron action potentials reach them with different latencies, contributing to the spatial gradation of muscular contraction. The twitch of a muscle as a unity caused by one action potential might, in the first approximation, be mathematically modeled as a critically damped system of the second order (Figure 2.10).[55]

$$F(t) = F_o \frac{t}{T} \cdot e^{-t/T} \qquad (2.11)$$

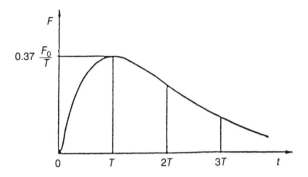

FIGURE 2.10 Idealized waveform of skeletal muscle twitch. (From Vodovnik, L. 1985. *Nevrokibernetika,* Ljubljana, Slovenija: Fakulteta za Elektrotehniko. With permission.)

where T is the time of reaching maximum force, about 75 ms in m. soleus. By increasing the frequency of neural action potentials to about 50 per second and more, muscle fibers no longer have time to relax and tetanic contraction takes place.

The gradation of the total force of a muscle as a unity is achieved by the activation of more motor units (spatial summation) and by increasing the frequency of the action potentials of particular motor units (temporal summation). According to the "motor unit size law," small motor units are activated first and then large ones. Hatze postulated the principle of "maximal grading sensitivity,"[62] according to which, contractile muscular activity is, in a certain sense, inverse to the activity of sensory organs-receptors described by the Weber-Fechner law. The total force (strain) manifested at the ends of the muscle in a longitudinal direction is the geometrical vector sum of the twitches of active motor units.

The relation between EMG signal and muscle force has been researched mostly in conditions of isometric contractions. Locomotion, however, is primarily characterized by dynamic contractions, in which the regimes of eccentric and concentric contractions are constantly exchanged. The activity of lower extremities musculature is of primary consideration. Skeletal muscle research in humans and animals has resulted in the establisment of their biomechanical characteristics which, at a macroscopic level, relate their kinematic properties (length, e.g., contraction velocity) with their ability to develop mechanical force (e.g., strain).

Force (strain) developed in a tetanically stimulated muscle in isometric conditions is a result of the length of its fibers. At a macroscopic level, the force-length relationship depends on the muscle as a whole (Figure 2.11). It is clear that maximum strain is achieved at the natural length of the muscle in the body. On this curve, the final increase in strain with length reflects contribution of serial elasticity concentrated in muscle fibers, tendons, and the remaining tissue.

Under conditions of anisometric contraction, the muscle is loaded with less or no resistance force and so, when motor units are activated, geometrical shortening occurs (concentric contraction). With no load, the contraction is isotonic which is an ideal state possible only in practice if the muscle is separated from the bone

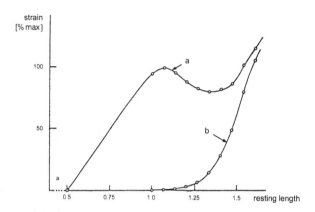

FIGURE 2.11   Idealized force-length relationship of the skeletal muscle in isometric contraction. (a) The total curve; (b) contribution of the serial elasticity concentrated in muscle fibers, tendons, and remaining tissue.

(experiments on animals). For tetanic stimulation, the following force-velocity equation is applied:[63]

$$v = b\frac{F_o - F}{F + a} \tag{2.12}$$

where: F = force of shortening
F_o = maximum isometric force with velocity = 0
a and b = empirical constants, specific for a particular muscle which is presented in Figure 2.12.

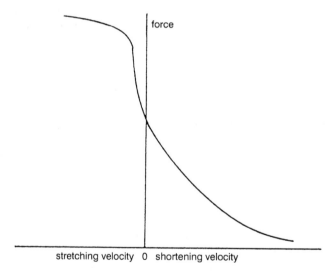

FIGURE 2.12   Idealized force-velocity relationship of the skeletal muscle: Hill's curve.

Eccentric contraction (see Figure 2.12), which is better called eccentric action (as Cavanagh suggested[64]), occurs when a muscle develops strain due to stimulation during which it therefore "tries" to contract, but external—more powerful—forces act in the direction of its lengthening. It has been estimated that metabolic energy can be most efficaciously converted into mechanical energy during contractions with an intensity which is 20 to 25% of the maximum and that it occurs at approximately 30% of the maximum contraction velocity.

The simplest way to represent skeletal muscle as a biological generator of mechanical force macroscopically, in the mechanical sense, is shown in Figure 2.13 which includes the active contractile element, the serial elastic element, and the parallel viscous-elastic element. The actions of the active contractile element, and partially that of the serial elastic element,[45] which is the dominant element shown in this figure, are directly related to the stimulation of a muscle via the motor neuron (see Figure 2.7), and thereby with EMG information, which is of special interest to us (Chapter 6). Those interested in the mathematical modeling of the muscle-tendon system (the cybernetic-biomechanical-biophysical approach) can find a systematic treatise on the subject in Zajac.[65]

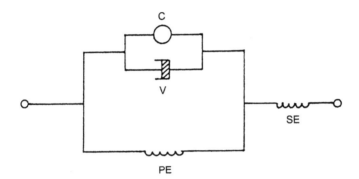

FIGURE 2.13   Equivalent mechanical scheme of the skeletal muscle. The symbols denote: C, active contractile element; SE, serial elastic element; V and PE, parallel viscous-elastic (attenuating) element.

3 General Properties of Locomotion Measurement Systems

This chapter presents what is common to most locomotion measurement systems: their global structure (Section 3.1), the presence of the analog-to-digital signal conversion procedure (Section 3.2), and the requirements imposed on systems by the user (Secton 3.3).

3.1 STRUCTURE OF A MEASUREMENT SYSTEM

In the simplest case, data obtained by measurement of human locomotion may be directly presented in a way suitable to the user, e.g., by plotting the time course of a measured variable. In general, however, "raw" measurement data have to be subjected to certain processing in order to be transformed into a form suitable for presentation and interpretation. The methods described in this book differ significantly with regard to the kinds of physical transformations of signals (as well as numerical-logical transformation of information), but a unique flow of information is established in all of them: starting with measurement, processing, and on to presentation, i.e., display of data.

Input data for the measurement system may be of varying physical character—a mechanical force, a geometrical quantity, a bioelectrical signal—depending upon which manifestation of locomotion is being measured. In general, the following process takes place: a measurement transducer transforms the input variable into a corresponding electrical quantity. Further possible processing is done on signals in their electrical form, and only the final data display appears yet again in some other form suitable for interpretation, e.g., through a plotter curve or a video display or via an alphanumeric record.

Figure 3.1 presents the general block schematics of a computer-supported measurement system. Chapters 4, 5, 6, and 7 cover the particular functional units of such measurement systems, primarily those concerned with measurement and digital data acquisition. The present chapter treats one subsystem of the presented schematics which is of utmost importance for the digital acquisition of practically all kinds of locomotion measurement data. This is the analog-to-digital (A/D) conversion of signals subsystem: the procedure by which, in general, a continuous time function is transformed into a sequence of digital numbers in order for it to be compatible with the digital computer. Both the instrument designers (engineers) as well as the

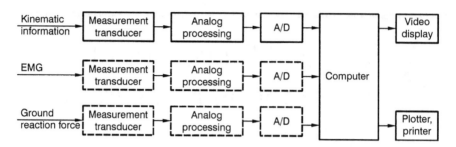

FIGURE 3.1 General block scheme of a computer-supported measurement system.

spectrum of their users (physicians, kinesiologists) have to be aware of the mathematical basis and the technical realization of the procedure.

According to some prognoses,[66,67] computers based on parallel processors will
probably become widely used. They will be particularly well suited to solving problems of the multisensor information type, pattern recognition, etc., i.e., applications
typical for the subject treated in this book. The representation here, however, presupposes classical digital computer schematics of a "von Neumann" sequential type,
commercially available.

3.2   ANALOG-TO-DIGITAL CONVERSION
OF SIGNALS

The analog-to-digital signal conversion procedure is a two-stage process: signal sampling and conversion of successive samples into digital numbers.

3.2.1   SAMPLING

During the sampling procedure, the raw continuous signal experiences changes in the
time and spectral domains. The mathematical basis for this procedure is presented
here.[68]

With a band-limited analog signal x(t), such as the one depicted in Figure 3.2a,
u(t) is a sampling function; with a sequence of rectangular pulses of the unity height,
a; width, b; and a period, Δ (sampling time), as shown in Figure 3.2b, the sampled
signal $x_u(t)$ is expressed as follows:

$$x_u(t) = x(t) \cdot u(t) \tag{3.1}$$

which is shown in Figure 3.2c.

The periodic sampling function u(t) can be expressed as a Fourier series:

$$u(t) = \sum_{n=-\infty}^{\infty} K_n \cdot e^{j\omega_u nt} \quad \text{with} \quad \omega = \frac{2\pi}{\Delta} \quad \text{and}$$

$$\tag{3.2}$$

$$K_n = \frac{1}{\Delta} \int_{-\frac{\Delta}{2}}^{\frac{\Delta}{2}} u(t) \cdot e^{-j\omega_u nt} \, dt = \frac{b \sin(n\omega_u b/2)}{\Delta n\omega_u b/2}$$

(complex Fourier series cofficients).

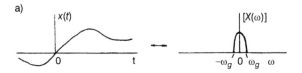

a)

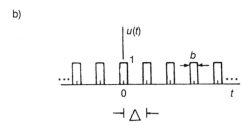

b)

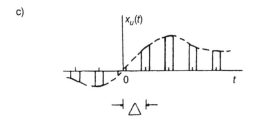

c)

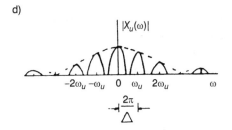

d)

FIGURE 3.2 Signals. (a) Analog signal in the domains of time and frequency; (b) sampling function; (c) sampled signal; (d) power density spectrum of the sampled signal.

By applying it to Equation 3.1,

$$x_u(t) = \sum_{n=-\infty}^{\infty} K_n \cdot x(t) \cdot e^{j \omega_u n t} \qquad (3.3)$$

Applying the Fourier transform,

$$F\{x_n(t)\} = x_u(\omega) = \int_{-\infty}^{\infty} x_u(t) \cdot e^{-j\omega t} \, dt \qquad (3.4)$$

($X_u(\omega)$ represents a power density spectrum), one obtains:

$$X_u(\omega) = \int_{-\infty}^{\infty} \{\sum_{n=-\infty}^{\infty} K_n \cdot x(t) \cdot e^{j\omega_u nt}\} \, e^{-j\omega t} \, dt$$

$$(3.5)$$

$$= \sum_{n=-\infty}^{\infty} K \cdot X(\omega - n\,\omega_u) = K\,X(\omega) + \sum_{n=-\infty}^{\infty} K_n\,X(\omega - n\omega_u)$$

$$n \neq 0$$

In the case of ideal sampling, with b = 0, the sampling function becomes a sequence of Dirac delta functions, and $K_n = 1/\Delta$, so the equation is transformed into:

$$X_u(\omega) = \frac{1}{\Delta} \sum_{n=-\infty}^{\infty} X(\omega - n\omega_u)$$

$$(3.6)$$

Hence, the power density spectrum of the sampled signal consists of a component directly proportional (factor K_o) to the power density spectrum of the original analog signal x(t) and of the sequence of additional spectral replicas, at distances ω_u, and multiplied by the corresponding factors K_n (Figure 3.2d). Using ideal low pass filtering, the original analog signal can be completely recovered from the sampled signal. The procedure is theoretically possible only if "overlapping" of spectra (aliasing) has not occurred, meaning that the sampling theorem must be valid:

$$\omega_u \geq 2\omega_g \quad \text{i.e.,} \quad \Delta \leq \frac{1}{2f_g}$$

$$(3.7)$$

Since an ideal filter cannot be realized, in practice it is necessary for ω_u to be even greater than the necessary minimum value, which is accomplished by combining analog low pass signal filtering (anti-aliasing) and increasing the sampling frequency f_u (practically up to $5f_g$). The sampled signal, shown in Figure 3.2c, might be denoted x_u (n Δ), since, ideally (b = 0), it attains values in discrete time intervals nT, while all other time is equal to zero.

3.2.2   CONVERSION OF THE SAMPLED SIGNAL

In the course of the analog-to-digital conversion procedure, a number of length r, expressed in parallel normal binary code (NBC), is attributed to each sample of the function x_u (n Δ). If we label the extrema of possible sample amplitudes (and the original signal x(t)) ± A, then the range of amplitudes of the input signal may be represented in L = 2^r discrete levels, with the constant quantization step d = 2 A/L. (Figure 3.3 illustrates the 3-bit converter transfer function.) Quantization noise is, therefore, inevitable; it may be labeled ε, and it equals the difference between the input signal amplitude and the closest discrete level. The value ε falls into the range ± d/2 and may be described as follows:

$$\varepsilon \, (n\,\Delta) = x_d \, (n\,\Delta) - x_u \, (n\,\Delta)$$

$$(3.8)$$

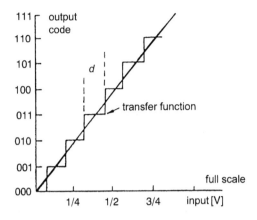

FIGURE 3.3   Transfer function of a 3-bit analog-to-digital converter.

where x_u (n Δ) denotes the real (accurate) value of the analog signal, x_d (n Δ) denotes the value of the digitized signal (the analog equivalent of the digitized signal), and ε (n Δ) denotes quantization noise. When estimating quantization noise with respect to useful signal, some assumptions regarding statistical properties of this quantity have to be introduced. Without entering into this procedure, the following relations may be presented.[69]

The signal to noise ratio (S/N) (the noise introduced by the quantization procedure) is

$$S/N = 20 \log (2^r \sqrt{12}) = 6.02r + 10 \text{ [dB]} \qquad (3.9)$$

The converter's dynamic range (the relationship between the maximum signal range and the quantization step d) is

$$D = 20 \log 2^r = 6.02 \, r \text{ [dB]} \qquad (3.10)$$

The equality between d and the analog RMS (root mean square) noise level in the original signal is considered optimal, which is the determinant of the maximum meaningful precision of conversion in a particular application.

3.2.3   TECHNICAL REALIZATION OF THE ANALOG-TO-DIGITAL CONVERTER

The state-of-the art solution to the A/D signal conversion procedure is in real time (on line). In past decades, however, a number of locomotion measurements were performed with measurement data stored as an analog magnetic record and were later (off-line) digitized and stored in a computer. The advantage of the procedure was the possibility of expanding the time scale when reproducing.

There are a number of methods for technically realizing the A/D converter. A method often used is the successive approximations procedure. An input analog signal is compared r times (r being amplitude resolution of the converter) to the internally generated analog signal. The internal signal is created by the digital-to-analog converter, so that its value in each iteration equals the sum of the previously determined approximate value of the input analog signal and one half of the remaining part of the measurement range. The partitions of the measurement range that are added are, hence, the following: $1/2, 1/4, 1/8, . . . , 1/2^r$. Results of the comparison (larger, smaller) determine the value of the tested bit and, correspondingly, also the value of an internal signal to be generated in the next iteration. The method is fast and efficacious, so modern hardware implemented converters attain conversion times in the region of a few tens of microseconds.

In locomotion studies, it is typical to simultaneously measure a number of signals which, accordingly, have to be subjected to the A/D conversion procedure. Due to the sequential functioning of the digital computer, it is possible to input only one datum to the bus at one time. Therefore, in order for the information from a number of measurement channels to be acquired in real time, the input subsystem has to provide time multiplexing of analog signals. The block schematics of the input subsystem shown in Figure 3.4 usually includes:

- An analog multiplexer (MUX)
- A "sample and hold" amplifier (S/H)
- An analog-to-digital converter (A/D)
- Logic for synchronization and control (LOGIC)

(It is presumed that input signals have been properly conditioned.)

Accordingly, within the defined time, the A/D converter provides the conversion of the signal value of those channels determined by control logic, controlled by the processor. There is a time lag between conversions of successive channels. The conversion procedure may also be initialized, if necessary, through external control logic.

In the Biomechanics Laboratory at the Faculty of Physical Education in Zagreb, one technical solution developed initially for the needs of industrial measurements was applied.[70] This was an 8-channel A/D conversion subsystem, of 8-bit precision (amplitude resolution), attaining a maximum sampling frequency of 1 kHz when all

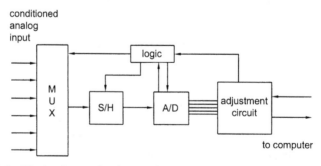

FIGURE 3.4   Block scheme of an input subsystem to a multichannel analog-to-digital converter.

8 channels were being measured, supported by an Apple II microcomputer. It was primarily applied to measure surface EMG and ground reaction force. The 8-bit amplitude resolution (256 levels), however, proved insufficient.[71] Another technical solution of superior precision (12-bit) and flexibility was developed subsequently, supported by the same microcomputer.[72] The 8 parallel channels were connected to a microcomputer via the subsystem (incorporating signal amplification through a Beckman Dynograph Recorder amplifier system, manufactured by Beckman, Inc. in the U.S.), while another 8 channels remained free for other measurement quantities. The software encompassed programs in assembler (assembly language) for multi-channel conversion. Conversion and digital acquisition procedures were guided by means of programs written in BASIC.[73] The minimum time needed between two subsequent readings was 46 μsec, and the maximum sampling frequency was 16 kHz while measuring only 1 channel. The user had the option to select measurement of only one or a paired number of channels. Selection of the sampling time was done through the keyboard, and technical accomplishment was based on software delay loop, written in assembler. An audible signal marked the beginning and the end of the conversion procedure. Data from RAM were later permanently stored on diskettes. A software package in BASIC enabled access to and manipulation of the acquired digital data, displayed by means of a cathode ray tube.

The solution described above was applied in a number of experimental locomotion studies, mostly in sports, and some of the results are presented in this book. Based on the hardware available commercially, the developed software support enabled a flexible realization of the measurement procedures and later access to the data for visual analysis and determination of conditions for the numerical processing of signals captured. In view of the kinematic, kinetic, and myoelectric signal features generated by human movement (Chapters 4, 5, and 6), time and amplitude resolutions of the system described were satisfactory. The system's main limitation is the fairly small memory capacity of the microcomputer used which, when measuring several signals ("raw" EMG), was only able to register measurement records of a second or two in duration.

With the development of the computer technology, however, the memory problem has been overcome. In the Biomechanics Laboratory, an A/D converter developed at the Department for Electronic Measurements and Information Processing at the Faculty of Electrical Engineering and Computing in Zagreb is used, making it possible to acquire measurement data on an IBM compatible PC. This is a data logger that enables the storage of relatively long (expressed in seconds), multichannel (a maximum of 16) measurement records with an overall maximum sampling frequency amounting to 333 kHz when 1 channel is being measured. Amplitude resolution is 12 bits, input dynamics is ± 5V, and the sequence of sampling input channels is programable. It works within a trigger regime and the number of samples, the triggering level, and the slope can be programed; especially convenient is the so-called pre-trigger regime mode. On-board dynamic memory is of 512 kB capacity with the possibility of enlargement to 1 MB. Software support enables the recording and storing of data, data inspection in the off-line mode, and cursor measurements in the domains of time and frequency, as well as zoom-in measurements.

Further, in electromyographic measurements, the technical solution applied is an A/D converter integrated into a Mega 3000 device (from Finland), enabling the conversion of up to 4 channels with a sampling frequency of up to 20 kHz (when 1 channel is being measured), with 12-bit amplitude resolution and the possibility of monitoring measurement signals in real time visually (on-line monitoring) via the PC monitor.

In general, it can be said that the specification of features and instructions for using an A/D conversion procedure in contemporary measurement systems is, in a way, analogous to instructions that were given earlier for "classical" measurement by means of various instruments (gain adjustment, adjustment of frequency characteristics of amplifiers using filters, etc.). The process of defining and choosing a desired mode of data acquisition and signal processing is greatly standardized and user-friendly.

3.3 REQUIREMENTS OF LOCOMOTION MEASUREMENT SYSTEMS

To be user-friendly, a measurement system (see Figure 3.1) and the corresponding measurement procedure protocol have to fulfill certain basic requirements:[3,74–76]

- Noninvasiveness, i.e., a decreasing degree of interference in the subject's performance of a movement: This is a classical requirement in the measurement procedure in experimental research fields while here noninvasiveness is specifically understood in an even more narrow sense than, for example, in medicine. Not only should a corresponding measurement sensor not physically enter the organism (for instance, as intramuscular electrodes do in clinical electromyography), but the measurement procedure, as such, should not significantly disturb a movement structure. This is especially critical when measuring fast and skilled movements. This requirement may also be more difficult to fulfill in elderly subjects or those with disturbed locomotion.
- Validity, meaning that precisely the quantity to be analyzed is being measured: This requirement is important in movement analysis and is not considered here from a merely technical standpoint (where the measurement is physically defined exactly, good and reliable data need to be acquired, which is understood intrinsically), but in view of the kinesiological interpretation, which may be considered as mutually independent requirements. If, for instance, EMG signals are used as the measure of muscular force, then it is wise to question the validity of this measure, since the relationship between muscular force and its bioelectric activity is not simple nor competely known at present. Furthermore, if we provide a certain measurement with the goal of evaluating the stability of human gait, then the term stability has to be defined precisely from the biomechanical standpoint. Consequently, this will determine the validity of the corresponding measurement method and of parameters chosen to evaluate gait stability. So, in general, the validity of the measurement procedure will be the result of the

purpose and properties of performance of a movement structure and of the choice of analytical (biomechanical, neurophysiological) model. The validity of a certain procedure or test is determined, in principle, in correlation to a criterion for the grading of properties measured which has already been accepted (clinical grading scale, expert knowledge of a gymnastics judge, etc.). Each new test should, therefore, have determined metric characteristics, calculated by standard statistical techniques.

- Reliability, requirements for accuracy and precision of the measurement procedure: Precision is determined by the physical-technical properties of the measurement method. For instance, force can be measured with defined precision expressed in mN. Accuracy may be determined with respect to the known corresponding reference values, which is the subject of defining the complete measurement procedure, i.e., the calibration of an instrument, the determination of body anthropometric approximation characteristics used, etc. This feature is verified by adequate statistical methods.
- Repeatability: Each measurement session should be able to be reproduced in a number of time instants (days), and the results obtained should correctly reflect the features of the phenomenon measured. The requirement of repeatability is also important when applying a certain measurement method, i.e., procedure, in different laboratory facilities.
- Possibility of direct, real time (on-line) data presentation: This feature gives the operator better and/or more efficacious control of the measurement procedure. This also allows information to be fed back to the subject during measurement, which may be advantageous in certain applications.
- Affordable price

There are additional requirements which may be imposed on kinematic measurement methods:[77]

- The possibility of absolute (in 3D) system calibration (which might be classified under "Reliability")
- Software to provide an inverse dynamic approach: This requirement is concordant with the "broader" meaning of the term of locomotion measurement adopted in this book, which includes procedures of data processing and interpretation as well as that of locomotion diagnostics. The software support for certain specific signal processing methods and for mathematical models for animation of movement, etc. may also be included here.

Locomotion measurement methods may be categorized into two principal, i.e., "clinical" groups: "functional-diagnostic" and "research." Clinical, i.e., functional-diagnostic, applications signify standardized measurements which are accomplished simply and quickly, with prompt feedback information, and are suitable for testing a large number of subjects in a short time. Research applications, on the contrary, are intended for realizing broader scientific investigations, which, in principle, do not require immediate feedback information. Each particular requirement mentioned

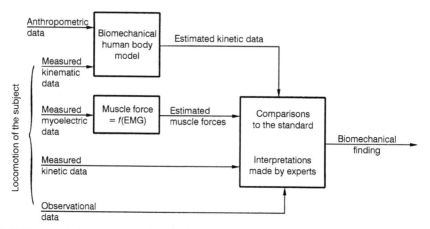

FIGURE 3.5   Information flow in a clinical locomotion measurement system.

above will, therefore, be adequately accentuated. The possibility of directly present-
ing and displaying data is, for instance, of special importance to clinical measurements and routine sports testing.

Finally, it should be stressed that a locomotion measurement system is not a
rigidly defined set-up, but a collection of technical devices and methodological procedures which are constantly being developed and improved. The general trend in the
past (which is also true for other fields of biomedical engineering and technology)
was the transfer of technical solutions initially developed for military purposes or
space research to areas of locomotion biomechanics. Similarly, the same is true for
certain other areas, such as the automobile industry (biomechanical mannequins). As
computers have become less expensive, however, original bioengineering applications have developed rapidly, e.g., the application of computer graphics within the
framework of the system mentioned earlier by Delp et al.[46] Figure 3.5 illustrates the
sequence of information in a locomotion measurement system.

4 Measurement of Locomotion Kinematics

Kinematics is the part of mechanics concerned with the description of movement. The term itself was coined in the mid-19th century by Ampère and stems from the Greek word κινεω, meaning to move.[33] Kinematic quantities represent an exact geometrical description of spatial (3D) movement of the body; kinematics may therefore be called the geometry of movement. The kinematic description of movement encompasses positions, velocities, and accelerations of body segments and angles, angular velocities, and angular accelerations between segments. Accordingly, input data to the measurement system are some of the quantities mentioned, depending on the actual method.

There is one thing common to all kinematic measurement methods: a necessary, *a priori* introduction of approximations. They are a result of the approach to the study of the phenomenon of locomotion and consist of the approximate determination (estimation of position) of particular characteristic body points. For instance, it is clear that it is not possible to locate the exact position of the center of the knee joint geometrically and mechanically because the knee itself is a rather complicated biological mechanism. For human movement analysis, however, it might, be appropriate to approximate knee joint geometry in certain instances by a simple one-axis mechanism, and accordingly, corresponding characteristic point(s) at the leg's surface, the landmark(s), whose kinematics should be followed, are determined. According to Section 2.1, the same is also true for many other geometrical and mechanical idealizations.

There are a plethora of kinematics measurement methods and in the last decade the field has developed greatly. In this chapter, methods have been classified into three groups: exoskeletal systems (Section 4.1), stereometric methods (Section 4.2), and accelerometry (Section 4.3). Stereometric methods are the largest and the most diverse group, therefore, the most coverage is devoted to them. These methods are crucial to other fields as well, such as industrial measurements, robotics, etc. The chapter ends with a description of processing methods for the kinematic data measured, aimed at minimizing errors in the measurement procedure and in providing an inverse dynamic approach (Section 4.4).

4.1 EXOSKELETAL SYSTEMS

The term exoskeletal system comprises both measurement and active "assistive" devices which, when attached to the body, function as an orthosis. Only exoskeletal measurement systems are considered here; and of these, only those applicable *in vivo.* A basic exoskeletal measurement method of this kind is electrogoniometry, standardly applied at the beginning of the 1970s and characterized by its simplicity and reasonable price.

Electrogoniometry means measuring angles between body segments. The technical devices used are called electrogoniometers. Typically, these devices use mechano-electrical transducers, rotational potentiometers. In the rotational potentiometer, electrical resistance changes in linear proportion to the angle of rotation of the axis. Therefore, if one supplies a constant voltage to the fixed connections (contacts) of the potentiometer, then the voltage between the movable connection (which follows the motion of an axis) and the fixed connection will be linearly proportional to the angle of rotation of an axis. The following equation describes this relation:

$$angle = constant \times voltage \tag{4.1}$$

The potentiometer's connections are fixed mechanically to straight "leads" which should be positioned parallel to the body segments between which one aims to measure the angle. (In the knee joint, for instance, "leads" need to be attached parallel to the upper and lower leg.) The goal is to position the potentiometer's axis within the imagined axis of rotation of the joint. The resulting electrical signal is an analog measure of the angle.

Angles between body segments are spatial, i.e., a complete determination of the angle comprises all three spatial dimensions, and the development of electrogoniometer devices has led to systems which use three mutually orthogonally positioned rotational potentiometers for each measured joint. These three potentiometers create a unity which is connected to corresponding units for the measurement of neighboring joints (hip-knee-ankle, for example). In such a design, which has to be adaptable and suitable for various sizes of extremities, the problem of unwanted sensitivity of the device to translatory movements in a joint exists (which always occur in reality). (The device is not capable, however, of measuring translatory movements.) Therefore, devices have been designed which use specially constructed connecting parts which by and large eliminate the influence of complex translatory movements on the measured rotation angle, since the electrogoniometer assembly "follows" the movement of an imagined center of rotation with respect to the joint as a whole. This kind of "self-aligning" electrogoniometers allow translational movements in a joint up to 4 to 5 cm without influencing the measured angle. By using elastic strips, the complete assembly is attached to the subject's body. The voltage supply, as well as the measurement signals, are transmitted by wire connections.

The accuracy and precision of these devices are determined by their mechanical design and by the electrical properties (linearity) of the potentiometer used. Chao[78] gives an example: the Duncan Pot R No. 1020–556, mechanical dimensions 12 mm in diameter, 14 mm in length, 1 K $\pm$ 10% resistance value, and linearity 1.0%.

The 3D Canadian CARS-UBC Electrogoniometer System has been used with healthy subjects and has proved suitable for the measurement of normal gait.[79] This low mass device (0.68 kg) has proven to be robust and accurate. In order to eliminate the influence of translatory movements, a so-called parallelogram chain mechanism was implemented in the instrument's construction. There were minor objections to the device's mechanical layout, but all in all, it was successfully applied together with the remaining gait measurement equipment—the footswitches—and measurement signals were digitized in real time and stored in the digital scope's memory. In fast locomotion, however, the device was not suitable since it oscillated mechanically, becoming invasive and inaccurate.

As an illustration, the results of electrogoniometric measurements of normal gait by Bajd et al.[80] are presented in Figure 4.1. The recorded time changes in the angles of the hip, knee, and ankle joints (in 1D) are shown, together with corresponding diagrams of foot-floor contacts. If a certain gait pathology is measured, this kind of record may serve as a useful addendum to the patient's clinical documentation.

This measurement of foot-floor contacts by means of appropriately designed switches is probably the most popular gait measurement method.[7] Combined with elementary instruments, such as a meter and a stopwatch, it provides the basic kinematic information on gait: average speed, rhythm, and step length. It is accomplished very easily, but it is relatively invasive. Measurements of this kind date back to Marey and Carlet in the 19th century (Section 1.2), and due to their high degree of simplicity, clinical usefulness, practicality, and above all, reasonable price, new solutions are being proposed.

Chao[78] and Chao and An[81] presented the principles of mathematical description of the spatial geometry of the knee joint in a subject measured by the electrogoniometer. Figure 4.2 shows the convention of a knee joint description in 3D according to these authors: the movements of flexion-extension, abduction-adduction, and internal (endo) rotation-external (exo) rotation are defined. This kind of description should be considered critically within the context of issues of geometrical (mechanical) modeling of the knee joint (Section 2.1).

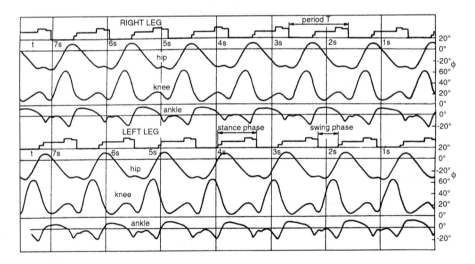

FIGURE 4.1   Hip, knee, and ankle angles, in 1D, as a function of time, in a healthy walking individual, measured by the electrogoniometer. (From Bajd, T., Kljajić, M., Trnkoczy, A., and Stanić, U., 1974. *Scand. J. Rehab. Med.* 6:78–80. With permission.)

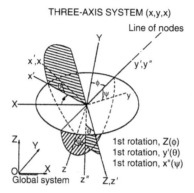

THREE-AXIS SYSTEM (x,y,x)

FIGURE 4.2 Convention for geometrical description of the spatial angle in the knee joint, similar to the gyroscopic system. (From Chao, E.V. 1978. *CRC Handbook of Engineering in Medicine and Biology, Section B: Instruments and Measurements,* Feinberg, B.N. and Fleming, D.G., Eds., Boca Raton, FL: CRC Press, 385–411. With permission.)

Electrogoniometers measure corresponding angles directly. Since the angles are often information which is needed for the analysis of the locomotion studied, this feature is advantageous. It is evident, however, that electrogoniometers are only capable of measuring angles in selected body joints and are not able to supply complete kinematic information. The inverse dynamic approach cannot be realized by starting with electrogoniometer-generated data since there is no absolute spatial reference of kinematic quantities available and since only relative movements between body segments are studied. These measurement devices, therefore, do not satisfy the requirements of a comprehensive modern biomechanical analysis of locomotion.

The transmission of voltage supply and measurement signals via wire (cable) connections is a disturbing factor in this measurement method. Therefore, telemetric instruments would potentially be a much better solution, as in other movement measurement systems (EMG).

Electrogoniometers are primarily suitable for measurements of slow locomotions in rehabilitation medicine, generally in clinical gait measurements and gait research. In this domain, they are a suitable and inexpensive instrument, able to provide measurements with satisfactory repeatability.

4.2   STEREOMETRIC METHODS

Stereometric methods offer a comprehensive solution to the measurement of kinematic quantities since they enable the reconstruction, in three spatial dimensions, of instantaneous positions of a moving point in a global coordinate system. As mentioned in Section 2.1, if this is done for at least three noncollinear points marking a certain body segment, then the position vector and the rotation matrix of this segment may be determined. Stereometric measurement methods of human movement may be classified in several ways.[35,75,78] The classification used here is according to Cappozzo,[35] and includes stereophotogrammetric, light scanning, and stereosonic

methods, with a modification to this last subgroup so that it encompasses a few inno-
vative systems operating on different physical principles.

An important contribution to the field comes from Herman J. Woltring
(1943–1992) from The Netherlands, who elaborated the methodology of stereometry in human locomotion measurements in a series of publications. Furthermore,
Woltring explained certain elements of semiconductor electronics in detail, which
were important for the design and layout of the commercial measurement method
SELSPOT, made in Sweden. By applying procedures from measurement theory, signal analysis and system identification, and signal processing to kinematic measurement systems and signals, he was able to give comprehensive and authoritative
criticism of their scientific validity and clinical applicability. Also, Woltring combined his knowledge of biodynamics, thereby establishing himself as one of the most
complete researchers of the biomechanics of human movement, experimental
methodology in particular. His considerations were supported by a detailed analysis
of the kinematic measurement device SELSPOT. In the late 1980s and early 1990s,
Woltring introduced internet communication to the field of biomechanics.

4.2.1 STEREOPHOTOGRAMMETRIC METHODS

4.2.1.1 Close-Range Analytical Photogrammetry Fundamentals

In 3D stereophotogrammetric measurement systems, recording and digital acquisition
of data are generally followed by photogrammetry. The mere recording and acquisition of kinematic data may be accomplished by various physical, i.e., engineering,
means and with varying degrees of accuracy, i.e., automatism, details of which will be
presented in the descriptions of particular measurement methods. Photogrammetry is
common to them all. In praxis, this procedure can be solved by using a number of
numerical methods, and the mathematical basis applied will be presented.

The problem in photogrammetry is in the spatial reconstruction of the location
of the observed object from convergent projections. (The beginnings of analytical
photogrammetry are found in Renaissance painting and the development of projective geometry.) By applying methods of analytical photogrammetry, the 3D coordinate values of a certain point P may be reconstructed with regard to the global
(external, object space) coordinate system (X, Y, Z) from coordinate projections of
this point on at least two planes located inside appropriately positioned cameras
(Figure 4.3). The image in each of the cameras is obtained by means of central projections of the global space on a light- (photo-) sensitive image plane through a system of lenses; however, the physical aspect of the procedure is not essential at the
moment. Projections are expressed in an internal coordinate system, or in the so-
called image space, (x,y,z), where z = 0 by definition, which is defined for each of
the two cameras. Equivalently, it is possible to combine three cameras, each with a
1D record,[82] which has been applied in some kinematic measurement systems.

In low-range photogrammetry, as in the study of human locomotion, the Eulerian
right rectangular (Cartesian) global coordinate system is suitable. Low-range photogrammetry comprises situations when the distance between the camera and the
object of interest amounts to no more than several hundred meters, typically about
1000 ft (308 m), so this method is appropriate to the study of human locomotion

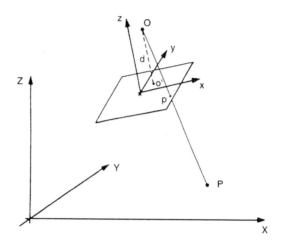

FIGURE 4.3 Stereophotogrammetry. The position of the measured point P within the global coordinate system and its projection onto the local coordinate system in one of the cameras. (From Cappozzo, A. 1985. *Biomechanics of Normal and Pathological Human Articulating Joints*, N. Berme, A.E. Engin, and K.M. Correia de Silva, Eds., Dordrecht, The Netherlands: Kluwer, 53–81. With kind permission from Kluwer Academic Publishers.)

measurements. Slam (1980, from the "Manual of Photogrammetry," according to Reference 83) defines photogrammetry as "the art, science, and technology of obtaining reliable information about physical objects and the environment through processes of reading, recording, and interpreting of photographic images and patterns of electromagnetic radiant energy and other phenomena." It is of crucial importance that the definition encompasses various physical measurement methods and not, as in the beginning, only the photographic. This analytical presentation is according to Gosh (1979, according to Reference 35) and Woltring and Huiskes[83] and is based on a two-camera setup. Each camera has a 2D internal record.

Prior to developing geometrical transformations as part of the photogrammetric procedure, several basic terms concerning camera construction and positioning in the recording field need to be defined. The camera consists of two main parts: the lens and the detector. These two parts are in a mutually fixed relationship, defined by the design and layout of the device itself. Lenses, crucial physical elements of a camera, may in general be classified as wide-angle (f_1 = 10 to 25 mm), normal (25 to 60 mm), and telephoto (60 to 200 mm). f_1 denotes the lens's focal length. F/stop marks the quantification of the lens aperture and is defined by the expression f_1/diameter of lens aperture. The field of focus (depth of field) is the range of distances within which objects appear in focus in an image. Focal length is determined by the lens equation:

$$1/f_1 = 1/a + 1/b \tag{4.2}$$

where a denotes the distance between the object and the lens and b denotes the distance between the lens and the film (picture), i.e., the location where the picture is in focus. The shutter factor is the part of the total circle of the circular shutter through which light passes onto film. The value ranges practically from 4(90°) to 3(120°) and may be changed. In order to realize good-quality measurements, image blur should be less than 0.05 mm.

External camera parameters are position, attitude, and focal length which may be adjusted from the outside. Internal camera parameters are position of the image space with respect to the optical axis and image distortion. The term "space resection" denotes estimation of external parameters from a set of control points. The term "camera calibration" denotes estimation of internal camera parameters from the same set of points. 3D position reconstruction of observed targets, however, is called space intersection. Ideally, in an error-free case (errors will be listed in Section 4.4.1 under errors of complete kinematic measurement procedure, whereas particular systems include software packages for error correction), the relationship between the position of the point in the image space and the position of the marker in the object space would be linear.

Let us assume the values of the internal coordinates of points measured are known and have been obtained by measurement using a camera, no matter in which manner, i.e., by any concrete measurement method, and reside in computer memory. The memory, therefore, has a sequence of successive values of kinematic displacement function. The instantaneous position of a point P in a global coordinate system is defined by Expression 2.1. Coordinates of this point are X, Y, and Z (see Figure 4.3). The corresponding internal point p in a certain camera is defined by coordinates x, y, and z with respect to internal space and is fixed to the projection plane in the camera. The center of perspective, determined by the system of lenses, is marked by O. Projection of this point on the plane x, y, z (principal plane), the point where the image plane and an optical axis cross, is denoted by o'. Their distance is marked by d (principal distance).

In the internal coordinate system, the vector which determines the position of point p with respect to the center of perspective is

$$[p - 0] = \mathbf{r} = \begin{bmatrix} x_p - x'_o \\ y_p - y'_o \\ 0 - d \end{bmatrix} \tag{4.3}$$

since $z_p = 0$, $z_o = d$, $x_o = x'_o$, $y_o = y'_o$ by definition.

In the external coordinate system, the following holds for point P:

$$[P - 0] = \mathbf{R} = \begin{bmatrix} X_p - X_o \\ Y_p - Y_o \\ Z_p - Z_o \end{bmatrix} \tag{4.4}$$

Rotating the axes of the global coordinate system, so that they become parallel to the axes of the internal local coordinate system, in the new coordinate system X', Y', Z', **R** is

$$\mathbf{R'} = \begin{bmatrix} X'_p - X'_o \\ Y'_p - Y'_o \\ Z'_p - Z'_o \end{bmatrix} = [M] \, \mathbf{R} \tag{4.5}$$

[M] is the rotation matrix of the local with respect to the global coordinate system. According to optical principles in producing images of P, vectors $\mathbf{r}$ and $\mathbf{R}'$ are collinear by definition. Therefore, after the axis rotation, the following expression holds:

$$\mathbf{R}' = k\,\mathbf{r} \tag{4.6}$$

From Equations 4.5 and 4.6, the following holds:

$$\begin{bmatrix} X_p - X_o \\ Y_p - Y_o \\ Z_p - Z_o \end{bmatrix} = k\,[M]^T \begin{bmatrix} x - x'_o \\ y - y'_o \\ 0 - d \end{bmatrix} \tag{4.7}$$

where k is a variable depending on the spatial position of point P. According to Woltring and Huiskes,[83] if $\mathbf{X}_p$ is in a given plane with known global coordinates, the constant k may be eliminated using the constraint equation:

$$\mathbf{N}'_o\,(\mathbf{X}_p - \mathbf{X}_o) = 0$$

where $\mathbf{X}_o$ is the global location of the defined point in the plane, and $\mathbf{N}_o$ is the vector normal with regard to the plane, with the solution:

$$\lambda = \mathbf{N}'_o\,(\mathbf{X}_o - \mathbf{X}_{co}) \,/\, \mathbf{N}'_o\,\mathbf{R}'_{co}\,(\mathbf{X}_p - \mathbf{X}_c) \tag{4.8}$$

This equation can also be expressed in the form of the direct linear transformation (DLT).[84]

$$\begin{aligned}
x_p &= (a_1 X_p + a_2 Y_p + a_3 Z_p + a_4)/D_p \\
y_p &= (a_5 X_p + a_6 Y_p + a_7 Z_p + a_8)/D_p \\
D_p &= (a_9 X_p + a_{10} Y_p + a_{11} Z_p + 1)
\end{aligned} \tag{4.9}$$

In this form, DLT is obtained after normalization with respect to the projection $\mathbf{X}_{co}$ onto the optical axis of the camera. Therefore, this factor must not disappear, which may be secured through the choice of origin of the global coordinate system somewhere within the monitored space. For the constant Z_p, the above equation describes general transformation of perspective between two planes (X,Y) and (x,y).

Expression 4.9 includes parameters which characterize the geometry of the camera used as well as input data: coordinates of a position in the image plane, no matter what the concrete technical solution may be. Camera geometry, as mentioned earlier, encompasses internal and external quantitative data. This expression depends on concrete measurement conditions, by which boundary conditions of the equation are defined. Due to the number of unknowns, at least two cameras are needed. Arrangements with more cameras are possible (3, 4, 5), which will be mentioned in presentations of particular measurement methods. Naturally, by increasing the number of cameras, both redundancy and reliability of the whole procedure are increased. The equations mentioned determine photogrammetric theory of the first order, since they describe an ideal theoretical case. The second order theory includes systematic errors in such a procedure, and they will be considered under errors in kinematic measurement methods (Section 4.4.1).

From an analytical viewpoint, it is useful to note that rotational (orientational) matrices consist of nine correlated elements which may be defined by various rotations. Possibilities are successive rotations around axis X with an angle ω, around Y with angle ϕ, and around Z with angle k.[85] So M becomes:

$$M(\phi) = K(k) \times \Phi(\phi) \times \Omega(\omega) \qquad (4.10)$$

with $\phi \approx (\omega, \phi, k)^T$ (^{T} denotes a transposed vector or matrix) and

$$K(k) = \begin{bmatrix} \cos k & \sin k & 0 \\ -\sin k & \cos k & 0 \\ 0 & 0 & 1 \end{bmatrix}$$

$$\Phi(\phi) = \begin{bmatrix} \cos \phi & 0 & -\sin \phi \\ 0 & 1 & 0 \\ \sin \phi & 0 & \cos \phi \end{bmatrix}$$

$$\Omega(\omega) = \begin{bmatrix} 1 & 0 & 0 \\ 0 & \cos \omega & \sin \omega \\ 0 & -\sin \omega & \cos \omega \end{bmatrix}$$

For pure orientation matrices M, $M^{-1} = M^T$.

4.2.1.2 The High-Speed Photography Method

The high-speed photography method was the dominant method until the 1970s, and is still often used in locomotion study, particularly in sports. The movement structure is recorded using two (or three or four, although theoretically only two are necessary) time-synchronized high-speed photographic cameras, usually 16- or 35-mm, usually with a speed within a 50 to 100 frames per second interval. At times, especially when a fast-movement structure is analyzed, velocities of up to 500/s are required. The kinematic signal frequency spectrum may reach 200 Hz during impulsive impacts. In general, the width of the frequency spectrum of kinematic signals is considerably lower, rarely reaching more than about 30 Hz. However, bearing in mind later processing of kinematic data by double differentiation, an ever higher speed of recording is needed which corresponds to the sampling frequency (Section 3.2.1).

The intermittent pin registered camera satisfies such requirements as it records at speeds up to 500/s. Typically, 25-mm lenses are used. Lens apertures are 22, 16, 11, 8, 5.6, 4, 2.8, and 2. Film sensitivity is 400 ASA. Prior to measuring movement, light intensity has to be measured.

Figure 4.4 illustrates a possible two-camera spatial arrangement for kinematic measurement. (The photographic method may also be applied to 3D measurements by

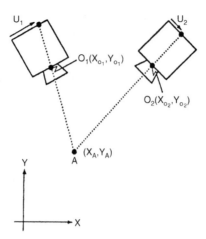

FIGURE 4.4   A two camera-spatial arrangement in a photographic stereometric, kinematic measurement method. (From Ladin, Z. 1995. in *Three-Dimensional Analysis of Human Movement*, P. Allard, A.F. Stokes, and J.-P. Blanchi, Eds. Champaign, IL: Human Kinetics, 3–17. With permission.)

using only one camera and by appropriately positioning mirrors, according to the method applied by Bernstein.) In practice, cameras are always positioned at a fixed angle: most often in line with and perpendicular to the line of the subject's movement. (Arrangements in which the subject is followed with a camera will not be considered here, although this is possible and is often a method of choice due to the higher spatial resolution achieved. Of course, it is then necessary to modify the photogrammetric expressions applied.) The location of cameras has to be accurately defined within a global coordinate system and must be such that their working space encompasses a part of the trajectory of the subject's movement which is to be analyzed. As mentioned in Chapter 3, in order to calibrate a system prior to measurement (and after), an object of known dimensions and location has to be recorded within a global coordinate system. This procedure enables the system to be calibrated within a global coordinate system. Its quality is the determinant of accuracy of the measurement procedure.

Once the recording procedure has been completed, film has to be developed. Next, the developed film record of a certain camera is projected, frame by frame, onto the monitor of a film analyzer-digitizer. Here, the coordinates of characteristic body points (x,y) are read. These points may be marked prior to measurement. If they have not been marked, the successfulness of the readout will depend exclusively on the operator and his knowledge of human anatomy. There are various engineering designs of the device for display and digitization of film recorded. So-called cross-hair systems, for example, have two movable, mutually perpendicular scales whose intersection has to be manually positioned in the determined point, after which values of coordinates are automatically recorded on a computer. The digitizer produced by the American company Lafayette has a resolution of 0.25 mm and absolute accuracy of ± 0.5 mm. Digitizers may also be constructed on the principle of ultrasound reflection. A chosen spot on the picture is touched by means of a manually controlled probe and corresponding values of coordinates are automatically transferred to computer memory. This procedure is repeated for all the remaining characteristic points in the picture. Then, everything is repeated for the next time instant (frame), etc. All this is repeated for recordings from the second or remaining cameras.

By applying the photogrammetric relations (Equation 4.9), from the extracted image coordinates (x,y for all of the cameras), global coordinates of characteristic body points (X,Y,Z) are calculated (see Figure 4.3). At the end of the procedure, the values of the calculated 3D coordinates of characteristic body points in consecutive time instants are stored in the memory.

Next, the spatial relations of characteristic body points, whose kinematics has been measured, and of anatomical body features have to be taken into account, as presented in Section 2.1. This depends on the layout of markers and their spatial positioning with regard to particular body segments, i.e., imagined joint centers. In this way, the trajectories of anatomical body points of interest are determined. Data so obtained may be typically displayed in the form of a stick diagram. The procedure is as follows: any angles or other relative spatial relations between particular body segments may be computed for consecutive time instants. A simpler, but often (clinically) useful possibility is the phase diagram which shows the interdependence of two chosen kinematic variables in a two-axes rectangular coordinate system, for instance, the angle in one joint and that in another.

Furthermore, by using numerical differentiation of displacement values, velocities may be computed, while further differentiation yields accelerations of particular body points, i.e., segments, which is also true for all the other stereometric kinematic measurement methods. These methods will be presented in Section 4.4.2. This enables the realization of the inverse dynamic procedure.

Some movement structures may, in a first approximation, be analyzed biomechanically in one plane only: for instance, the long jump in a sagittal plane. Then, basic kinematic information is 2D: one camera positioned perpendicularly to the sagital plane suffices for measurement and so the mere procedures of measurement and calculation are simpler and less time consuming. However, in principle, with regard to the complexity of the human body, at least two cameras are needed for high-quality measurements and analyses.

The California Group, mentioned in Section 1.2 (the Berkeley Group), first took measurements in 1944. Several kinematic measurement methods were used: photographic, accelerometric, photographic with mirrors, and other biomechanical measurement methods such as electromyography and ground reaction force measurement. The high-speed photography technique, with 16-mm film and a speed of 30/s, was applied in a 3D kinematic measurement method, however, not in concordance with the exact analytical photogrammetric procedure, but via a simpler yet similar procedure. Three orthogonally positioned cameras were used: laterally, frontally, and transversally from above with regard to gait direction. Volunteers participated. Pins were used as markers and inserted transcutaneously into the femur and tibia (!). Their projection was deduced from each "view." Changes in the rotation of these projections with regard to directions observed in referent anatomical position and attitude were interpreted as angles of flexion-extension, abduction-adduction, and endorotation-exorotation of leg segments, while differences of corresponding angles for the tibia and femur were interpreted as joint angles.

Invasiveness was the major problem of this method. Woltring also criticized the described approach since, although deviations of segment angles from real values in such a method were not large, angles in the joint deviated considerably. During flexions of large amplitude in the femur and tibia, as well as during swing phases of gait, it becomes critical. Further, calculated joint angles depend on the attitude of both segments with respect to the camera, even if there is no relative movement between them[86,87] (Inman et al. 1981, according to Reference 31). Cappozzo and Paul[88] point to the error caused by the requirement of holding one hand still while walking. The research by the California Group nevertheless produced six so-called kinematic determinants of normal gait which, in summary, determine the trajectory of the center of gravity of the body during gait.

"Compass gait" is the initial and most simple case of a bipedal system in which, one imagines, there is only flexion and extension in a hip joint. Lower extremities are represented by rigid levers without foot, ankle, and knee mechanisms. The pathway of the center of gravity of the body is described by a series of intersecting arcs whose amplitude is considerably greater than that in normal gait (Figure 4.5). Angular

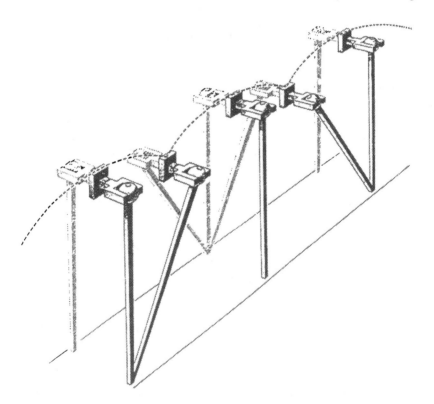

FIGURE 4.5 Compass gait: a crude approximation of the model of human gait kinematics, accomplished only with hip joint rotations. (From Saunders, M., Inman, V., and Eberhart, H.D. 1953. *J. Bone Jt. Surg.* 35-A(3):543–558. With permission.)

rotation in the hip during flexion is equal to the same during extension. A healthy subject can intentionally simulate this crude approximation of gait by trying to walk on his heels with fixed, extended knees, which appears very "clumsy" visually and is inefficient energetically.

The first gait determinant that is added to "compass" gait is pelvic rotation (Figure 4.6). Normally, it reaches a total amplitude of around 6 to 8 degrees with respect to the direction of progress. This additional degree of freedom contributes to a smoother trajectory of the body's center of gravity, more harmonious transitions between successive "arcs" in "compass gait," and so a decrease in energy consumption. Meanwhile, angular rotations in the hip decrease. A similar technique is used by walking racers.

The second determinant is pelvic tilt (Figure 4.7). In normal gait, the pelvis is tilted downward relative to the horizontal plane on the side opposite to that of the weight-bearing limb (positive Trendelenburg). To enable pelvic tilt, the swing leg must be flexed in the knee joint. The lowering of the swing hip occurs rather abruptly at the end of the double support phase. The trajectory of the center of gravity is lowered further, the pelvic trajectory becomes smoother, and by flexing the knee of the swinging leg, energy is conserved by the effective shortening of the pendulum.

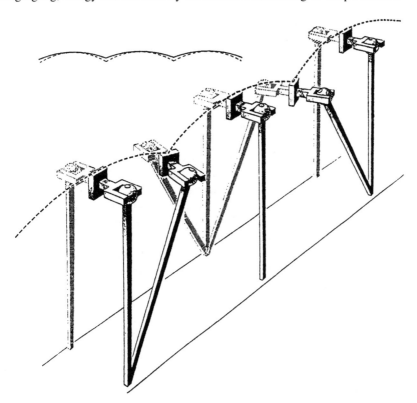

FIGURE 4.6 The first gait determinant which, in addition to compass gait, includes pelvic rotation. This mode of walking resembles the one observed in race walkers. (From Saunders, M., Inman, V., and Eberhart, H.D. 1953. *J. Bone Jt. Surg.* 35-A(3):543–558. With permission.)

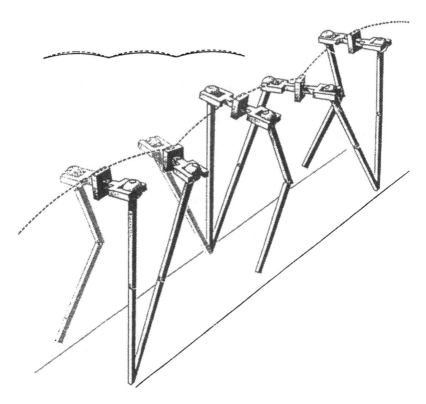

FIGURE 4.7   The second gait determinant, including pelvic tilt. (From Saunders, M., Inman, V., and Eberhart, H.D. 1953. *J. Bone Jt. Surg.* 35-A(3):543–558. With permission.)

The third gait determinant is knee flexion in the stance leg (Figure 4.8). It always occurs in natural gait. On average, the flexion angle is around 15 degrees. The gait model becomes ever more natural, approaching a situation where there are actually two flexions of the stance leg, of which the second occurs prior (or coincidentally) to the heel being lifted off the ground.[87] The amplitude of oscillations of the center of gravity increases further, as does energy expenditure.

The fourth and fifth gait determinants encompass foot and knee mechanisms. The problem is in smoothing out the pathway of the center of gravity in the plane of progression at the point where its arcs intersect and where, as previously illustrated, there are inflection points. During the stance phase, two intersecting arcs of rotation are established (Figure 4.9). The first arc occurs during heel contact and is described by the rotation of the ankle around a radius formed by the heel. The second arc is made by the rotation of the foot around a center established at the front part of the foot in association with heel rise. Prior to toe-off, a plantar flexion of the foot begins, synchronically with the beginning of knee flexion. Saunders et al.[87] describe the kinematics of corresponding mechanisms in ankle and knee joints and their combined influence which results in the smoothing out of abrupt inflections at points where the curves, described by arcs, intersect and which show a forward progression of the center of gravity.

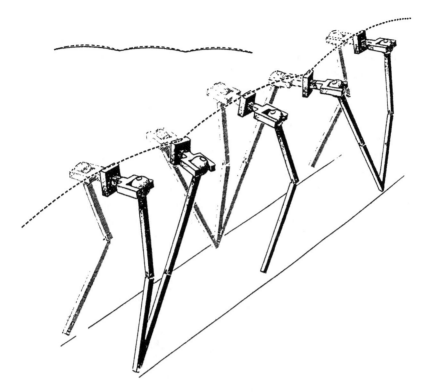

FIGURE 4.8   The third gait determinant, also including the stance leg knee flexion. (From Saunders, M., Inman, V., and Eberhart, H.D. 1953. *J. Bone Jt. Surg.* 35-A(3):543–558. With permission.)

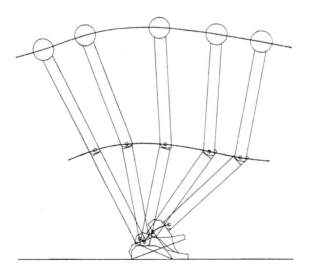

FIGURE 4.9   The fourth and fifth gait determinants: foot and knee mechanisms. (From Saunders, M., Inman, V., and Eberhart, H.D. 1953. *J. Bone Jt. Surg.* 35-A(3):543–558. With permission.)

Finally, the sixth determinant is lateral displacement of the pelvis. During gait, the center of gravity of the body is displaced laterally over the weight-bearing leg twice during the cycle of motion. This is accomplished by the horizontal shift of the pelvis or by relative adduction in the hip. Due to the tibiofemoral angle, excessive lateral displacement is corrected (Figure 4.10). All in all, the effect achieved is that deviation of the center of gravity is almost symmetrical in both horizontal and vertical planes. This results in a smooth and natural gait trajectory (Figure 4.11).

Each gait determinant mentioned is generally a function of one degree of freedom in one joint, and they have been listed to include an ever greater number of degrees of freedom of movement, by which real gait is more and more realistically modeled. By incorporating them, energy expenditure becomes ever more rational (just the first determinant means an estimated 300% increase in energy expenditure relative to normal gait). The determinants listed are generally considered to be a clear

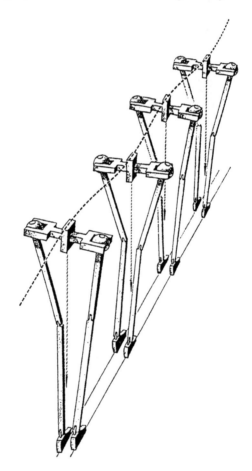

FIGURE 4.10 The sixth gait determinant, also including lateral pelvic displacement. Correction of excessive lateral displacement is achieved through the influence of the tibiofemoral angle and adduction in the hip joint. (From Saunders, M., Inman, V., and Eberhart, H.D. 1953. *J. Bone Jt. Surg.* 35-A(3):543–558. With permission.)

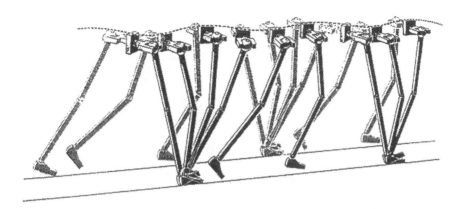

FIGURE 4.11   A summary effect of several gait determinants to the trajectory of the body's center of mass. (From Saunders, M., Inman, V., and Eberhart, H.D. 1953. *J. Bone Jt. Surg.* 35-A(3):543–558. With permission.)

kinematic gait model. Different pathologies may be analyzed with respect to the model shown, and in each of them, locomotor function adaptation may be understood as the organism's attempt to minimize energy expenditure needed for locomotion. Five out of six gait determinants lower vertical displacement of the center of gravity by which net energy expenditure (cost) is decreased, as are other qualities of this locomotion; this has been further confirmed.[89]

In a later development in the California Group, Sutherland and Hagy[86] applied a similar, yet entirely noninvasive measurement method by filming gait with 16-mm cameras positioned at both sides of the walking path and in the frontal plane, with a speed of 10/s. To illustrate, results of applying this kind of measurement procedure on a group of eight subjects are shown (four children older than seven and four adults) (Figure 4.12). The average curve in the group is shown, as well as the ranges for flexion-extension movement in the knee joint. Of the results shown, this one shows the least intrasubject scatter, i.e., the largest degree of repeatability in the group.

Of all the investigations mentioned—historical and present-day—the photographic method has been applied in a great number of kinematic measurements of healthy and pathological locomotions alike. Atha[74] systematically classified photographic techniques applicable to human motion analysis, beginning from the elementary procedure of a single photographic shot (record) to complex 3D solutions of cinephotography. The diversity of this technique is therefore evident. This classification supplements what has already been said in Section 1.2.

1. One camera recording, one shot
   - Action photography
   - Photography with pose
2. One camera recording, more pictures
   - Stationary film, open shutter, low ambient illumination
     –Time exposures: enables "blurred" recordings giving the illusion of movement
     –Cyclography

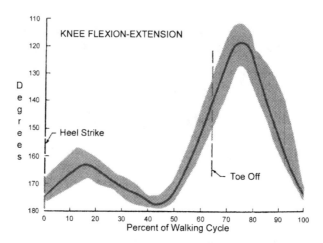

FIGURE 4.12 Average value and range of time curves of the flexion-extension angle in the knee joint in eight healthy subjects. Abscissa is marked in percentages of the gait cycle duration. (From Sutherland, D.H. and Hagy, J.L. 1972. *J. Bone Jt. Surg.* 54-A(4): 787–797. With permission.)

–Chronocyclography
–Stroboscopy: a field of view is intermittently illuminated by means of regular low-frequency (5 to 30 Hz) light impulses (multiple exposition)
• Moving film in front of an open shutter
–Streak-cyclography, streak-chronocyclography
–"Stripped"-photography
• Alternating framed film
–Time-lapse cinephotography
–Serial photography
–Standard cinephotography
–Slow motion cinephotography
–High-speed cinephotography
3. Multi-camera recording
• 3D recording
• Stereophotography

In modern biomechanical analyses of human movement, the methods of high-speed cinephotography with one camera and multi-camera methods are of primary interest.

Besides attaining a very high spatial resolution and accuracy of about 1%, the basic quality of the photographic method is its noninvasiveness, i.e., not disturbing the subject, which results in a high practicability of application in various field situations, such as sporting, competitions, including high-level events, such as the Olympics, etc. (This, of course, does not include the invasive method by Saunders et al.[87]) A further, important comparative advantage of this method is permanent storage of the complete photographic record of the movement structure recorded. This offers an "informationally rich" background (context) during analysis, as well as a possibility to redefine characteristic body points in later analyses. This is further suitable for a combined qualitative and quantitative analysis.

The basic drawback of the high-speed photography method, which is extremely off-line since one needs to develop the film, is the need for manual frame digitalization, a very slow and cumbersome procedure, prone to human error. Greaves[90] gave an example of the digitization of a measurement record of body movements during the performance of a golf swing. If one records with four cameras, with the chosen model of the body defined by 17 markers, and if, for instance, a speed of 200/s is used, then up to 7 hours of constant "clicking" by the specified mechanism are needed for the analysis of a sequence lasting no more than 2 seconds, with an operator speed of one reading per second ($17 \times 4 \times 2 \times 200 = 27,200$ data points)! Apart from the procedure being very slow, under these conditions, reliability and repeatability of the procedure become questionable, since the operator tires significantly. The method is also not suitable for direct application in real time (on-line) because there is a time lapse between filming and data presentation and application of results. Also, error correction during measurement is not possible. In principle, this method is therefore not suitable for clinical medical applications or for fast, diagnostic sports testing. High-speed photography may initially be useful for the measurement of transient phenomena, i.e., unique cycles of happenings, with a later (off-line) results analysis. Cavanagh,[91] however, cited the work of Kasvand et al. (1971) as being an attempt to scan the film record automatically. Ingels et al. (1969) showed the precise analytical stereophotogrammetric method for measurement of 3D coordinates, but with application to the dynamics of the heart muscle.

To conclude, the value of photography is that it stores a great quantity of information on movement and the human eye may later detect important details from the record. Many laboratories use film in the early phases of new projects. A significant drawback to this method is the inability to correct the experimental protocol during actual measurement.

What has been considered so far refers to classical photography. The time may come when the classical photochemical procedure used in photography will perhaps be completely replaced by direct electronic storage of light information. If this happens, the classical photography method as a method for measuring locomotion will become obsolete.

4.2.1.3 Optoelectronic Methods

In the photographic method, after the film has been developed (a considerable delay in itself), the process of analysis and digitization, frame by frame, is slow and cumbersome and prone to human error. In an effort to circumvent this procedure, methods and procedures have been developed that are based on direct conversion of visual geometric information into an electronic record of a kind that forms the basis of automation of the measurement process. These stereophotogrammetric methods may be labeled optoelectronic. The systems appeared in the late 1960s as specific laboratory solutions to begin with and later as commercial products. As material and solid state physics, along with microelectronic technology as well as consumer electronics in general developed quickly after the 1960s, solutions were of better quality and at a more reasonable price. In the past, expert engineering knowledge was required for realization and practical application of systems, while today they have become commercially accessible, at reasonable price, are user friendly to a great extent, and are more flexible in application. This trend will surely continue, with significant impact from the field of computer sciences (computer vision, pattern recognition, and artificial intelligence). With regard to the interesting and dynamic development of this field, some of the pioneering optoelectronic methods for measuring human movement kinematics, followed by presentation of those currently in use, are presented.

Furnée[92] gave an acceptable classification of optoelectronic measurement methods for the kinematics of human movement:

1. TV camera-based, i.e., video-kinematic systems
2. Systems based on position-sensitive analog planar sensors
3. Systems based on 1D array of sensors
4. Systems based on mechano-optical scanners

The systems will be described accordingly, with one modification: Group 4 in the classification list will be described in Section 4.2.1.4, Light Scanning-Based Methods, according to Cappozzo.[35]

Classification could be based on some other criterion as well. For instance, the systems could be classified as those with active and passive cameras or active and passive markers. Mere classification, however, is not so crucial. What is important is that the development of the idea and the technical realization of the procedure of automatic detection of body markers is conveyed to the reader.

Method 1. Video-Kinematic Systems: These are the earliest systems. The image coordinates of passive reflective noncoded body markers or small light bulbs were automatically read by means of a custom-designed interface and transferred to a recorder or computer for further processing[93] (Furnée 1967, according to Reference 30). Marker image coordinate values (x,y in Figure 4.3) were deduced from the detected marker position in a TV camera scan. Identification of corresponding marker image points in different video frames (in time) was made by means of pattern recognition software, usually via some form of extrapolative prediction and closest neighbor search.[30] These generalizations do not refer, however, to frame-grabber type devices that also belong to this group. Common to all system solutions described is that they represent a measurement application of one standard product of consumer electronics that is not originally designed for measurement, but for human visual perception of stored, i.e., transmitted, information presented in a form of an image.

Historical Development of Automated Sensing Procedure in Video Kinematic Systems

In this section, the development of pioneering video kinematic measurement systems will be presented. During development, many principles of signal, i.e., TV image, processing became apparent and have remained, with modifications, almost until today.

Origins of Video Kinematic Measurement Systems

Furnée[16] referred to designs that may be considered to be the beginning of video kinematic locomotion measurement systems.

Wisnieff (1961) developed an automatic tracker in the form of an analog video processor for 2D coordinates of a single object (marker). A pair of gated electronic integrators gave voltages proportional to planar coordinates of a small bright target, to which the system was locked previously by manual windowing. Data (a contrast picture) were from the domain of military surveillance of missiles or aircrafts.

d'Ombrain et al. (1966) realized digitization of a simple stationary TV picture, a single-valued graph. The stationary picture was monitored by a slow-scan TV at 9 frames/s. Video output was graded in three levels of gray for recording on binary tape which allowed 250-byte spaces per TV line. The extraction of digital coordinates was done later (off-line) by computer processing the tape. By providing storage for the quantized video from every pixel, the procedure may, in effect, be considered a kind of early frame-grabber and not digitization in real time.

Schuck (1966) is mentioned by Furnée as being the first to realize real time digital coordinate extraction from the video signal, again, however, of a single object and the application was in tracking stars. A reduced TV scan rate of 3 frames/s was used, while several successive coordinate pair values were averaged. Digital processing circuitry derived the image coordinates of the monitored star by comparing the time base of the pulse (derived from the video threshold crossing) with the position of the vidicon's electron beam, determined by horizontal and vertical sweep voltages. (The way to track a vidicon tube beam was not explained.) The principle was used first in astronomy in a case mentioned by Janssen (1876), who used a photographic gun for photographic recording of the planet Venus before Marey (Section 1.2). Here again, the astronomer's application was concerned with a single object and slow motion in the field of vision, like Janssen who introduced time-lapse photography, as opposed to Marey's physiological study at high-speed recording and minimal aperture times.

The Quantimet Image Analysing Computer (Fisher 1967), an apparatus by Cambridge Instruments Ltd., Great Britain, which basically served for particle counting and measurements in metallurgy, but also in cytometry, enabled feature extraction from stationary pictures. An adjustable threshold was used, and the video was quantized to a binary black/white pattern. Pulse duration corresponded to the width of a white segment intersected by the TV line, and the determination of the sum of particle areas was through gated integration of these pulses. Furnée described the characteristics of this instrument in detail, but stressed the difference in its purpose which was counting rather than determining positional information.

Murphy et al. (1966) from the General Electric Company developed the Metrocamera. The function of the instrument, consisting of a TV camera and a block of logic circuitry, was to track an optical point object and provide the x,y coordinates of the point, at a speed of 60 frames/s and with a basic accuracy of 1:250. The mechanism was not described completely. If provided with active switched markers on the body, such a camera system might also be used in a multipoint mode.

Development of the Principles of Automatic Marker Detection and Threshold Suppression

At Delft University of Technology in The Netherlands, the development of a kinematic measurement system based on the TV camera, called Mark I, began in 1967.[16,93] (Mark I is also the name of one of the first digital computers by Howard H. Aiken at Harvard University in 1944.) Basic technology used commercially available monochromatic TV cameras with a vidicon tube, connected to an IBM-1130 minicomputer through a corresponding system for conversion, reduction, and preprocessing of data. (In 1968 a DMA interface prototype was developed.) Reflective markers, fixed to characteristic body points, were tracked. The usual arrangement included clusters of several noncolinearly positioned, mutually fixed markers on each body segment.

The essence of this automated measurement procedure was the system for conversion, reduction, and preprocessing of data which provided detection of the brightness threshold. The so-called video-digital coordinate converter used two electronic binary counters: an x-counter for the horizontal and an y-counter for the vertical image coordinate. A parallel readout register was linked to the x-counter (Figure 4.13). The counters were coupled to TV synchronization impulses, standard for horizontal and vertical synchronization. The horizontal coordinate was obtained from the counter which was incremented by a high-frequency clock. The position of any one point of an image, chosen on the basis of the brightness threshold crossing, was digitized directly into binary code by triggering the x-register, followed by readout of that register and the value y (y was incremented only when the video signal was suppressed and therefore did not require a separate register). Triggering occurred whenever the criterion for threshold detection was satisfied.

Each TV line is scanned, in the horizontal direction, during a time interval of 20 ms divided by 312 (number of lines), amounting to 64 µs, also including a standard 12 µs for return of the horizontal scan (European system). During the return period, the video signal is suppressed by the horizontal suppression impulse. Each horizontal

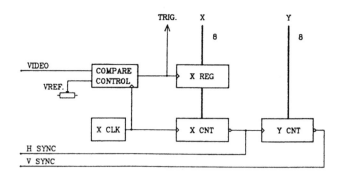

FIGURE 4.13   Block diagram of the video-digital converter of coordinates of a light marker in the Delft TV kinematic measurement system. (From Furnée, E.H. 1989. Dissertation. TV/Computer Motion Analysis Systems: The First Two Decades. Delft, The Netherlands: Delft University of Technology. With permission.)

scan (TV line) is initiated by the horizontal synchronization impulse which occurs during the suppression impulse. The horizontal coordinate of any picture element was thus a continuous value, defined by the value of voltage or current deflection. Such sawtooth voltage was determined by the passage of time after the horizontal synchronization impulse. Hence, the problem to be solved was reduced to measurement of time, in real time. The digital coordinate converter introduced discretization. Furnée (1961) refered to the original application of digital counting of the passage of time on a TV scan in radar.

In the 1967 prototype, an interlaced scan was used, comprised of two half-images: odd (1, 3, 5, 7, etc.) and even (2, 4, 6, etc.). This TV picture scanning mode is adapted to the inertia of the human eye, realizing a representation which is subjectively experienced without flicker, since there are two dark intervals during one picture. But, for measurement purposes, such a subjective, perceptive criterion is of secondary importance because first and foremost are accuracy and reliability of the measurement procedure. The interlaced scan, however, was used only in the prototype from 1967. Later, externally synchronized cameras were used with synchronization impulses derived from the system. Not distinguishing odd from even images, a noninterlaced scan was used.

Measurement of time was realized by using a 4-MHz system clock and corresponding 8-bit binary counters. There were 256 horizontal increments. Effective subdivision of the line of an image was 224. The horizontal synchronization impulse resets the x-counter, so that each TV line was divided into equal increments. As the relation in dimensions of a standard TV image was 4:3, the width to height relation of an imaginary pixel, defined as a grid 224:256, differs from 1. In the prototype of the 1967 solution, software correction was made which became standard in later systems with increased resolution of primary counters. Both nonlinearities, horizontal and vertical, were temperature dependent, so the system required 1 hour of warm-up prior to use. Also, the 1967 prototype device was the only one which used a TV camera in the normal mode (free running) and therefore used the synchronization impulses from the camera, mixed with the video signal, in correspondence with CCIR norms.

Since markers were the points of strong light contrast with respect to background, it was a 1-bit representation (monochromatic video). Only detection of transition from black to white was used in 1967. Beginning in 1970, both directions were used by application of univibrator circuits, i.e., clock-controlled digital differentiators.

As the marker image usually occupied two or more successive lines, conversion at the black-white threshold generated more coordinate pairs in image space (x,y), which corresponded to the upper contour of the respective marker. Therefore, to allow detection of neighboring markers present in a scene, after the first detection one had to define a certain rejection zone. It was realized by means of the so-called contour suppressor which inhibited readout of following contours (on the next line, etc.) into the mentioned zone. This was first accomplished by a cascade of bistable univibrators which gave suppression impulses of corresponding width τ and delayed by $(T - \tau)$, where T denotes period of TV line (Figure 4.14). This inhibited (suppressed)

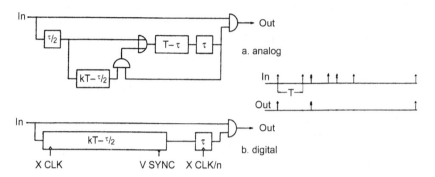

FIGURE 4.14   Analog (a) and digital (b) circuit layout for body marker edge supression in the Delft system. (From Furnée, E.H. 1989. Dissertation. TV/Computer Motion Analysis Systems: The First Two Decades. Delft, The Netherlands: Delft University of Technology. With permission.)

the conversion of coordinates of all points, except the upper one, at any continuous contour. In this way a greater number of vertically displaced markers could be accommodated. For horizontally displaced markers, on the same line, one such circuit per marker was needed. With improvement of the procedure, digital design was introduced. The main concept is shown in Figure 4.14.b, a shift register. This allowed a larger number of neighboring markers in a horizontal direction, since each was represented by several bits entering the 256-bit register. Black-white or white-black transition, or both, could be used. From the original method of determining only the left marker contour, which introduced error in unequal marker intensities, one moved over several options (Figure 4.15). Moving to the HP 1000 computer, at the beginning of 1980s, a procedure of computing the marker centroid was accomplished which determined a high resolution of the method. In this way, the procedure of automatic detection of marker centers was completed. In all stages of this system's development, the procedure took place in real-time. FORTRAN and assembler language were mainly used in the realization.

Transition to CCD (charge-couple device) matrix sensor-equipped cameras introduced many advantages. There were no more nonlinearities as a consequence of analog vidicon system, output became inherently digital, image smear was eliminated, and device warm-up was no longer necessary. Due to the production process of CCD sensor manufacture, characterized by accuracy, linearity, and stability,

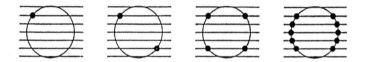

FIGURE 4.15   The options for determining marker edges in the Delft system. (From Furnée, E.H. 1989. Dissertation. TV/Computer Motion Analysis Systems: The First Two Decades. Delft, The Netherlands: Delft University of Technology. With permission.)

geometrical distortion was practically eliminated. Other semiconductor sensors like the MOS (metal oxide semiconductor) matrix sensor are also characterized by such properties. What remained after this was mostly due to lens errors.

Inspired by the prototype of the Delft University system, shown at the 7th Congress of the ICMBE in 1967, Jarett et al.[94] (at the Bioengineering Unit of Stratchylde University, Glasgow, Great Britain) developed a system through a cooperating project. The system used passive markers, covered by reflective paper (Scotchlite®). It allowed up to 6 TV cameras to be connected to the DEC PDP-12 computer, with a 16-K, 12-bit working memory, with peripheral memory at the magnetic tape and, later, a hard magnetic disk.

Markers were illuminated with a stroboscope and their reflections were detected by the simple threshold detection circuit (Figure 4.16). This was possible because of the high brightness so realized and also the insensitivity achieved to changes of spatial orientation of markers. Voltage threshold value was adaptive, taken as a proportion of peak value of voltage received. Peak value was continuously stored and refreshed by the marker signal. Changes in signal level caused by changes in distance from the marker during measurement were compensated in this way.

Jarett and colleagues[94] described engineering solutions considered for the layout of this system. A dedicated layout of computer interface was chosen with a DMA controller, representing one of the main differences from the Delft system. A 16-word buffer register was used, so that DMA transfer during fly-back was enabled, when there were no video-derived coordinate inputs. In this way, the number of markers at each individual line was limited to 5, since, during 12 μs of fly-back, only 7 words, 5 horizontal coordinates, 1 common vertical coordinate, and 1 additional status code word could be transfered. In cases where there were more than one marker per line, this scheme enabled a reduction of vertical coordinates to be transferred and stored. Compared to the Delft system, the advantage of this method of buffering was that markers were mutually close, while unbuffered DMA transfer in the Delft prototype determined a minimum distance of 1/16 line width due to transfer cycle time.

The Stratchylde system prototype did not have a marker contour suppression mechanism, but small markers were selected that practically occupied up to two lines. When generating more pairs of coordinates per marker, values were software averaged. When a certain TV line was crossed with the marker, one impulse of marker

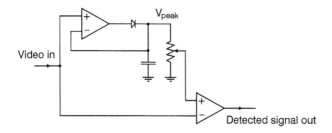

FIGURE 4.16 Marker detector circuit in the Stratchylde kinematic measurement system. (From Jarett, M.O., Andrews, B.J., and Paul, J.P. 1976. *IERE Conf. Proc.* 34:357–370. With permission.)

detection was generated, which caused reading of an x-counter. This impulse happened on the rising crossing of the video signal threshold and, accordingly, corresponded to the black-white (leading) edge of the marker image, while the width of the marker image, i.e., white-black edge, was not taken into account in this design.

Besides the advantages already discussed, an adaptive level of video threshold also introduced a degree of unpredictability in situations when all markers were not evenly lighted, such as in the case of nonplanar work. Cameras in this system were used in an interlaced scan mode.

In a newer configuration, a marker contour suppression mechanism was implemented, but as opposed to the Delft system, it detected the marker's leading contour and generated only one pair of coordinates per marker. A stroboscopic illuminator was developed with a xenon ring illuminator and later with rings of IR LEDs, with the goal of eliminating line-appearing noise in a vidicon image and realization of simultaneous sampling. In 1980 the PDP 12 computer was replaced by the PDA 11/34. This system could include up to 64 markers.

One of the first reported applications of TV cameras in human motion study is the one by Dinn et al. (1970) called CINTEL (Computer INterface for TELevision, according to Winter et al.[95]) at the Medical School of the University of Manitoba, Winnipeg, Canada. A CDC 1700 computer was used. Prior to measurement, the subject had to be marked with reflective markers. Halves of table tennis balls were used. One had to position larger reflective markers in the background as spatial reference points. Movement recording was on a tape at 60 frames/s, according to the standard American TV system. The tape was later scanned in real time, row by row, and the video signal was automatically continuously digitized so that the conversion of the analog to the digital signal was done on the basis of the brightness threshold (1, bright; 0, dark). This resulted in a data transfer speed of 20,000 words per second, if data were packed 24 bits to a word. One completely scanned image resulted in a matrix of zeros and ones, 96 × 96 in size. This was the "window" which encompassed all markers and fetched every second image line (interlaced mode). This conversion of a previously stored record was a precursor to the frame-grabber in real time.

The television-computer interface used was a general-purpose digital system which performed continuous A/D conversion of the video signal. It was an adjustable low pass video filter and a 6-MHz maximum, 5-bit flash A/D converter (1 bit was used here). Horizontal tact was adjusted to 6, 3, 2, 1.2, 0.75, and 0.5 MHz to give basic sampling impulses. Horizontal and vertical synchronization impulses were extracted from an internally synchronized camera and used to control the aperture for viewing and conversion. The aperture could be adjusted in a range from 1 × 1 to 256 × 256. A limiting factor with computers of that time was data throughput into internal memory, so only a limited number of CINTEL's possibilities in time and amplitude resolution and window size could be used.

Since body markers in this application were 4 cm in diameter, and the background ones were even larger, during scanning, each marker resulted in a number of elements of a binary matrix. Figure 4.17 shows printout of one such converted TV field. The center of the marker was determined indirectly, using numerical interpolation, by

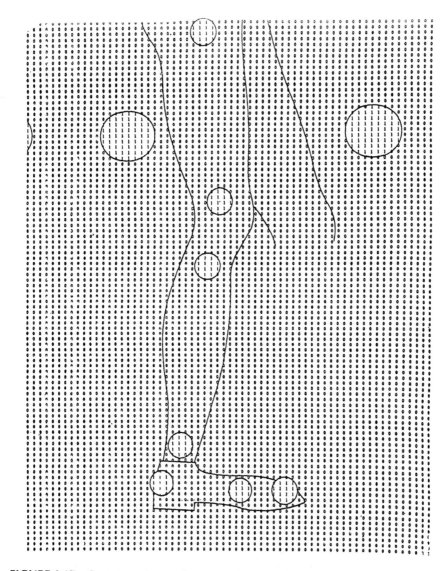

FIGURE 4.17 Computer printout of a converted TV field in the kinematic measurement system CINTEL. The resolution is 1-bit (0 for dark, 1 for light). The silhouette of the leg is plotted to indicate marker location sites. Larger markers in the background are easily identified and serve as absolute reference points in the plane of progression during walking. (From Winter, D.A., Greenlaw, R.K., and Hobson, D.A. 1972. *Comput. Biomed. Res.* 5:498–504. With permission.)

calculating the mean value of the coordinates of all points of a marker. Figure 4.18 shows the influence of marker size on the accuracy of the calculation of the marker center. With increasing marker size, the accuracy of this kind of estimation increased too, so in cases where the marker encompassed, for instance, 10 points, it was no

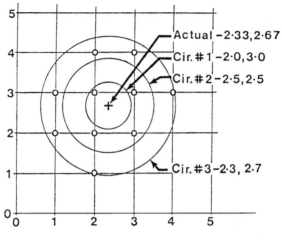

Spatial Resolution vs. Marker Diameter

FIGURE 4.18 Influence of marker size on the accuracy of calculating its center in the CIN-TEL system. (From Winter, D.A., Greenlaw, R.K., and Hobson, D.A., 1972. *Comput. Biomed. Res.* 5:498–504. With permission.)

longer necessary to sample each TV line to realize a wanted resolution of 1/10 distance between TV lines. Spatial resolution realized in this manner amounted to 1 mm (excellent for the needs of human kinematics measurements).

In this application, one camera tracked a walking subject from a 3-m distance. This was, naturally, reflected in the trigonometric relations of the applied stereometric relations which differed from those in Section 4.2.1. 2D global coordinates were calculated. The system was used for normal gait recordings, up to five strides in practice. The authors mentioned the possibility of including one more camera which should be positioned frontally with regard to the walking direction. In this way, the system would be potentially ready for measurement and capturing 3D kinematics.

Figure 4.19 shows time diagrams of measured displacements of characteristic body landmarks for two strides in a healthy subject.

In another application (Cheng et al. 1974, according to Reference 94), the TV camera was connected to the PDP 11/10 minicomputer. Markers were active and realized in the form of penlights. Specificity of this method was that scanning of the row was provided only for detection of the first marker, further coordinate generation was inhibited, and then one went on to the next row. This system was limited to the detection of only one marker per row. An internally synchronized 60-Hz TV camera functioned in an interlaced mode, so that coordinate data might vary by ± 1 LSB (least significant bit). The scanned image was 240 lines in size with 237 pixels, so 8 bit counters for x and y were used, and for each marker detection spike, a 16-bit word was latched into the register and presented to the host's UNIBUS. Data transfer was led by the program and data evaluation was provided by means of real-time software to detect errors. Realized horizontal resolution amounted to 1:241 and vertical to 1:246.

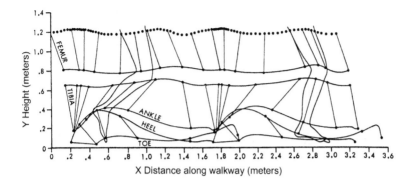

FIGURE 4.19   A typical printout of measured sagittal displacement values of body markers as a function of time, in a healthy walking subject. Two complete strides are plotted. The marker on the tuberosity of the greater trochander is shown for each second sample; the other markers are each 200 ms (each 12th TV field). (From Winter, D.A., Greenlaw, R.K., and Hobson, D.A. 1972. *Comput. Biomed. Res.* 5:498–504. With permission.)

Development of Marker Identification Procedures

After detection, the next problem to be solved was marker identification: a problem of all contemporary methods with nonaddressable markers. Namely, readout of TV format values is always the same from top to bottom (y) and from left to right (x). So, if two markers mutually exchange positions between two frames, for example, so that the smaller vertical coordinate in the first frame becomes the greater in the second (Figure 4.20), a change in the order of values of stored marker coordinates in the computer memory will take place (and so also in their identification). Therefore, a procedure has to be implemented in the software which will return the two values to their correct position. The principle according to which this can be accomplished is by considering values of successive coordinates that are supposed to form a trajectory—unique and continuous, possibly. Furnée described the development of corresponding algorithms in which linear and quadratic extrapolation were applied, principles of function monotonity were followed, or else *a priori* knowledge was used, including work by Steilberg (1968), van Ingen Schenau (1973), Nieukerke (1974), Antwerpen (1983), Demper (1985), and Zuidhof (1989).

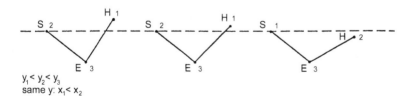

FIGURE 4.20   A disturbance in the order of storing extracted marker coordinates caused by the defined order of TV image scanning. The coordinates of markers 1 and 2 will exchange positions. Example of hand movement. (From Furnée, E.H. 1989. Dissertation. TV/Computer Motion Analysis System: The First Two Decades. Delft, The Netherlands: Delft University of Technology. With permission.)

Marker identification in the Stratchylde method followed a scheme of linear extrapolation (Steilberg 1968 and van Ingen Schenau 1973, according to Reference 16). Initially, one had to manually classify two sequential groups of data. Two-point 2D linear extrapolation was conducted according to the equation:

$$x\ (n + 2) = x(n) + 2[x\ (n + 1) - x\ (n)] \qquad (4.11)$$

Estimation of the mid-field point was then calculated by simple averaging in x and y directions. Then, after low pass filtering, a correction of nonlinearities was provided. Higher-order extrapolation for marker identification as well as further software for smoothing of measurement curves and estimation of derivative values were according to Andrews 1982 (quoted according to Reference 16).

The Delft and Stratchylde systems represented the origins of the development of the first devices commercially available for kinematics measurements, PRIMAS and VICON, respectively, which will be presented later. The TV Delft system used one camera, with up to 16 markers for each sample, but used only 5 of them in practice. The system described was applied to measuring hand motion in problems concerning prosthesis control with which the Prosthetics Control Lab Group at Delft University of Technology was occupied (M.J. Wijnschenk and E.H. Furnée). Most of the basic characteristics of the Delft system, such as software support for marker identification, remained in later versions.

Recent Video Kinematic Measurement Systems

Recent video kinematic measurement system solutions are based on cameras, with CCD or MOS matrix sensors, interfaced with computers. The typical methods for the automated marker recognition in this type of systems are described in this section.

Body Markers

Circular, disc-shaped, or spherical reflective (passive) markers are customarily used. For example, in the PRIMAS System (PRecision IMage Analysis System), by the Dutch firm HCS Vision Technology B.V., Eindhoven, which is a continuation of the Delft system, the markers are made of 3M Scotch® material, introduced to the field by the Stratchylde group.[16,93,96] They are laid out in several sizes. The standard practical solution has groups of three markers forming a cluster. Further, in the VICON (VIdeo CONvertor for biomechanics) System by Oxford Metrics Limited, Oxford, England, which is the first commercial kinematic measurement system dating from 1981, and a continuation of the mentioned Stratchylde system, spherical reflective markers 19 mm × 14 mm in size are used. Next, reflective semi-spherical markers could also be used, as is the case in the ELITE (ELaboratore di Immagini Televisive) System, manufactured by Bioengineering Technology, Milano, Italy, developed during the 1980s in the Centro di Bioinginieria di Politechnico di Milano, Italy,[15,97] since they can easily be applied to the body, and their image in a camera does not change if they rotate around the axis of symmetry. In this system, in measurements of gait, with a 7-m subject-to-camera distance, the best practical solution was shown to be a marker 5 mm in diameter (the span of practical layouts is from 1 mm to 1 cm).

During measurement, markers can be stroboscopically illuminated by means of a group of IR emitting diodes (infrared light emitting diode, IRLED, assembly). Pulsed IRLEDs are positioned circularly around a camera lens; these are the directed rings of LEDs, by which a directivity of both LED emission and of the reflecting groups of markers are compensated, so that maximally equal image intensities are realized in the total working area. This contributes to video contours being independent of position and gives a sharp image and simultaneous equidistant sampling of all markers. (When cameras are mutually in each other's viewing field, the strobes are slightly offset in time to prevent a camera from seeing an opposing illuminator.) There are more versions of illuminator layout, in concordance with the camera field of view, so as to realize an approximately homogenous intensity of marker images. In VICON, for example, a ring of IRLEDs is made around the lens of each camera which give a short (2-ms) light flash at the end of each video scan. In this way, image blur caused by fast-moving markers is avoided. To minimize the influence of background light, each lens is equipped with an IR low pass filter.

Cameras

One possibility is to use CCD TV cameras with an electronically controlled shutter and a stroboscope. (The PRIMAS System uses HCS MXR 100 dedicated 100-Hz (120 in the U.S.) (increased speed) cameras with an electronically controlled shutter and a stroboscope.) There is another option, valid generally in TV methods, of scanning only one half of the height of display at a frequency of 200 (240) Hz, whereby time resolution is increased, but at the expense of a smaller working space. Work may be carried out in daylight. The level of black is stable and there is no a pixel clock ripple. Image time integration is lowered and amounts to 0.1 ms (synchronized to 0.1 ms IR strobe) or 1% of image cycle time. By suppressing the component of constant scene illumination, the feature of double stroboscopy significantly enhances marker contrast and surmounts previous limitations caused by ambient illumination and background light. Applying an electronic shutter, a smear-free image of a moving marker is realized (Figure 4.21).

Cameras may also be synchronized by phase locked loop (ELITE). Fast cameras may be used, at the expense of decreased spatial resolution, and speeds of 50/60 up to 240 Hz are possible.

FIGURE 4.21 An IR camera equipped with a stroboscope comprised of LEDs, part of the Italian ELITE system (ELITE, commercial material).

The Principle of Threshold Detection

By keeping abreast of advancements in integrated circuit technology, the principle and layout of automatic conversion of video information in digital coordinates at the edges of marker images have not practically changed since the beginning and the hardware realized in principle in 1967 (the PRIMAS example) (see Figures 4.13 and 4.14). This is basically a count of TV lines and the number of pixels in the basic image framework and an instantaneous readout of these binary counters at video threshold crossings. These are, then, x and y coordinates in image space of detected marker contour.

Data reduction can be provided through estimation in real time of coordinates of marker centroids from digitized marker contours, enabling data transfer to a personal computer. This kind of procedure of suppression of marker contour has replaced the one in the Delft system. The centroid processor realizes the following algorithm (Figure 4.22):

$$x_m = [\Sigma_i \, Px_{mi}] / 2 \, \Sigma_i \, Dx_{mi}$$

$$y_m = [\Sigma_i \, Py_{mi}] / 2 \, \Sigma_i \, DX_{mi} \qquad (4.12)$$

with:

$$Px_{mi} = Sx_{mi} \, Dx_{mi}$$
$$Py_{mi} = y_{mi} \, Dx_{mi}$$
$$Sx_{mi} = (x_{1mi} + x_{2mi})$$
$$dx_{mi} = x_{2mi} - x_{1mi}$$

where m denotes marker index, i denotes index of marker segment line, and x_1 and x_2 are x coordinates of rising and falling marker image contour. Furnée[16] described the circuit implementation of these equations in detail. If marker size occupies 10 lines in an image, the substitution of 20 pairs of coordinates with one pair of values of centroids represents a significant saving for processor time. In this procedure, x and y values are defined by 15-bit precision, i.e., noise is very small. The procedure finally

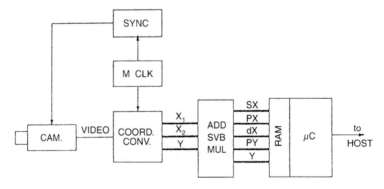

FIGURE 4.22 PRIMAS, block scheme of the circuit for coordinate calculation of the body marker center. (From Furnée, E.H. 1989. Dissertation. TV/Computer Motion Analysis Systems: The First Two Decades. Delft, The Netherlands: Delft University of Technology. With permission.)

results in the definition of markers with sub-pixel precision. Besides generating centroids, individual heights and widths of markers are generated as well. They may be used for checking the shape integrity of projected marker images later.

This, together with absolute synchronization of low-noise cameras to the system of conversion, gives very high frame-to-frame repeatability or precision of this kinematic measurement chain: 1:18,000 for x and 1:13,500 for the y coordinate. Time resolution is 100 Hz (European system), i.e., in the realm of additional possibilities mentioned earlier. Measurement noise is low, typically 0.11 mm of standard deviation for values x and y, in a working space 2 × 1.5 m in size. CCD technology enabled rigorous synchronization of the camera to the coordinate converter, up to the pixel clock level, which is the amount used for optimal stability of conversion.

Another principle of marker detection (VICON) is to apply the threshold to the image and store locations of points where it has been reached.[98] Manipulation of the threshold causes contours of all markers to be stored. In fact, in the first version of the commercial system, only the rising edge was detected, as in the Stratchylde system. A hardware method for threshold suppression did not exist: each one of the multiple images detected for each marker was resolved, software-wise, into a unique coordinate value. A separate 16-bit input register existed for controling sampling frequency as a submultiple of 50 images/s, as well as a simulated video test pulse generator. The principle for setting up the automatic threshold, which depends on video peak pulse history (Stratchylde), was abandoned and control of detection level was done manually. In VICON, the method was expanded to complete marker edge.

This method was a compromise between practical, low data rates and keeping as much useful information as possible. Hardware was redesigned so that symmetry of detection of left and right marker edges was improved, giving a more faithful reproduction of marker shape, and thresholding was redesigned to minimize flicker. Coordinates of marker edges were generated and stored in real time. Markers typically cover 4 TV lines, in which case it is necessary to generate a total of 8 pairs of horizontal and vertical coordinates per marker. Horizontal coordinates are generated from 1 left to 1023 right for 50 Hz. Resolution of coordinate generator, therefore, surpasses the number of pixels per line in the imager (then 800).

Vertical coordinates are generated from top to bottom. There are 250 to 300 lines per field, depending on camera standard. If cameras are used in noninterlaced mode, the vertical coordinate is determined by the ordinal number of the TV line multiplied by 2. If cameras in interlaced mode are used, odd images yield odd vertical coordinates, while even yield even. In other words, neither vertical jitter nor interlaced-induced noise are introduced.

2D centers of circular marker edges are obtained later (in the off-line mode). The software package called AMASS uses iterative fitting of a circle for this method. The determination of the center of the circle improves vertical and horizontal resolution significantly. Raw data are kept and can, if necessary, be processed again. Figure 4.23 shows the procedure of iterative fitting of a circle on two examples, where one sees the influence of horizontal resolution on procedure accuracy.[98]

Cameras functioning in the interlaced scan mode are described as more accurate, as opposed to conventional viewpoints. This is acceptable in the case of static measurements. If one measures static markers, results from both fields may be combined,

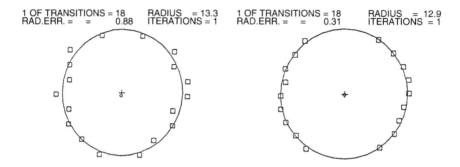

FIGURE 4.23   Principle for determining the marker center in the VICON system by means of circle fitting. (From Macleod, A., Morris, J.R.W., and Lyster, M. 1990. *Close-Range Photogrammetry Meets Machine Vision,* A. Gruen and E. Baltsavias, Eds., SPIE Vol. 1395, 12–17. With permission.)

and so the number of points on the circle used for calculation of a particular marker's center is doubled. It is especially important for linearization and calibration procedures in the 3D measurement procedure. Since static markers are used in calibration, their odd and even half images may be combined, doubling the vertical resolution of cameras (Figure 4.24).

In one commercial solution, a patented, quadrature edge detection was used (Expert Vision, manufactured by Motion Analysis Corporation, Santa Rosa, California).[99]

The Principle of Marker Shape Recognition

As opposed to previously described solutions based on the detection of the light threshold, automatic detection based on shape recognition is also possible (ELITE). Markers are automatically recognized by means of the dedicated computer algorithm by a pattern recognition technique for object identification in real time and according to shape and size, not exclusively to light intensity as is the case in previously described systems. In this way, the system allows greater freedom with regard to

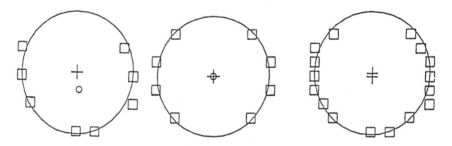

FIGURE 4.24   Increase in the precision for determining the marker center in the VICON system in the static situation in the interlaced mode operating regimen. (From Macleod, A., Morris, J.R.W., and Lyster, M. 1990. *Close-Range Photogrammetry Meets Machine Vision,* A. Gruen and E. Baltsavias, Eds., SPIE Vol. 1395, 12–17. With permission.)

lighting conditions and distribution of background illumination and is also applicable in sunlight.

Video signal processing architecture is realized at two hierarchical levels (Figure 4.25):[97]

- The dedicated peripheral fast processor for shape recognition (FPSR), designed and accomplished by means of a fast digital VLSI chip
- The general purpose computer

The FPSR is connected to a unit called ITE (Interface to the Environment) and calculates the cross-correlation of an input signal, previously digitized, and a stored mask (template). It thereby recognizes markers and calculates their coordinates, by which approximately 1000-fold data reduction is accomplished. As marker recognition is realized by using information on shape, a shutter is needed to avoid nonsimultaneous information readout by the electronic beam of a TV camera. The shutter acts as a stroboscope, leaving the lens open for about 1.5 ms, and is realized as a disk with transparent slim opening on a dark surface. The disk is controlled by a synchronous motor powered by the sinusoidal current, in phase with the vertical synchronism of the TV camera. The sinusoidal and vertical synchronism signals are generated in the FPSR.

Each 20-ms image is considered and an analog TV signal is sent to the FPSR for processing in real time. Here, digitization is provided at a sampling frequency of 5 MHz, which corresponds to later pixel matrix dimensions of 256×256, with 16 gray levels (4 bits). A chosen template which describes a marker is stored in the CPU. The shape detecting algorithm (SDA) is based on the 2D cross-correlation mentioned between digitized image and previously defined shape, dimensions 6×6 pixels. Only if and when the value of cross-correlation surpasses the previously determined threshold value are pixel coordinates digitized. If the threshold is surpassed and this does not correspond to marker image, e.g., as in the case of clutter, etc., it is rejected by this mechanism. In the 1986 solution, calculation of marker centroids in real time

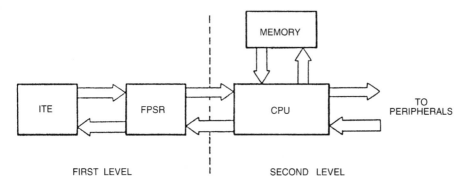

FIGURE 4.25 Two hierarchical levels of video signal processing in the ELITE system. (From Ferrigno, G. and Pedotti, A. 1985. *IEEE Trans. Biomed. Eng.* 32:943–950. With permission.)

did not yet exist and all coordinates of accepted pixels were dumped on the host computer, together with corresponding values of calculated cross-correlations.

The FPSR is implemented as a parallel hardware structure enabling real time processing. The output from the FPSR directly makes "r" pairs of horizontal and vertical coordinates of the "r" markers detected, which are sent to the CPU (level 2).

The SDA is described as follows. The TV image may be represented in 2D by denoting pixels with their row and column indexes, i.e.,

$$P_{11}, P_{12}, \ldots, P_{1N}, \ldots, P_{21}, \ldots, P_{ij}, \ldots, P_{MN}$$

A matrix of M × N pixels is therefore defined. The total TV signal contains both the marker shape and the background scene which is considered to be noise in the present analysis.

Defining:

A (x,y) = marker shape
A'(x,y) = correlation function representing previously chosen marker
          shape (template)
T (x,y) = total TV signal
F (x,y) = background signal

total signal is

$$T(x,y) = A(x,y) + F(x,y) \tag{4.13}$$

under the condition:

$$F(x,y) = 0$$

where A(x,y) # 0. Cross-correlation between functions T(x,y) and A'(x,y) is

$$R_{A'.T}(h,k) = \int\int_{-\infty}^{\infty} A'(x,y)\, T(x+h, y+k)\, dx\, dy \tag{4.14}$$

By substituting Equation 4.13 in Equation 4.14, one gets

$$R_{A'T} = R_{A'.A} + R_{A'F} \tag{4.15}$$

where

$$R_{A'.A} = \int\int_{-\infty}^{\infty} A'(x,y)\, A(x+h, y+k)\, dx\, dy$$

and

$$R_{A'.F} = \int\int_{\infty}^{\infty} A'(x,y)\, F(x+h), y+k)\, dx\, dy$$

For a limited dimension of the shape to be recognized,[97] corresponding discrete forms of these expressions are

$$R_{A',T}(h,k) = \sum_{0m}^{M'} \sum_{0n}^{N'} A'(m,n) \, T(m + h, n + k) \qquad (4.16)$$

and

$$R_{A',A} = \sum_{0m}^{M'} \sum_{0n}^{N'} A'(m,n) \, A(m + h, n + k)$$

$$R_{A',F} = \sum_{0m}^{M'} \sum_{0n}^{N'} A'(m,n) \, F(m + h, n + k)$$

where M' and N' represent width and height of chosen marker shape to be detected. The chosen shape, therefore, consists of $M' \times N'$ pixels, with $M' < M$ and $N' < N$. The problem is reduced to determining the function $A'(x,y)$ which maximizes the term $R_{A'A}$, i.e., A' must have low correlation with F and high correlation with A. Ferrigno and Pedotti[97] described the details of this procedure.

To increase resolution, calculation of the centroid of the detected marker is provided, i.e., of the centroid of points that surpass the threshold. A look-up table is used here to accept marker candidates and to calculate their centroid. Instead of pixel intensity (x_i,y_i), for the calculation of coordinate centroids, the values R_{ij} of a 2D cross-correlation with respect to the chosen marker template 6×6 in size are used in the following expressions:

$$x_c = \frac{\sum_i x_i \sum_j R_{ij}}{\sum_{ij} R_{ij}} \qquad y_c = \frac{\sum_j y_j \sum_i R_{ij}}{\sum_{ij} R_{ij}} \qquad (4.17)$$

Because this principle is based on recognition of defined marker shape, and not their brightness, the reliability of the procedure is very high. Furthermore, because all levels of gray are used in an image, real resolution is significantly increased. In this system, however, the size of the monitored markers is critical and only limited marker movement along the line of vision of the camera can be tolerated. This would limit the device's use to approximately planar movements monitored in a sagittal plane. However, markers are small, even smaller than 1/256th of the viewing filed, which in practice is projected to a size smaller than the defined marker (subtemplate). The necessary number of pixels may be covered by intentional blooming and defocusing of such images. (The authors consider this procedure to be a more reliable way of getting a subpixel resolution than by estimating geometrical centroids.) Due to the processing described, the marker centroid is calculated with a resolution of 1:65,536 of the field of view. The final accuracy, which includes distorsion corrections across the whole viewing field, is experimentally determined to amount to 1/2800 of the viewing field. In Ferrigno and Pedotti,[97] a resolution test was described on the basis of minimal discernible linear marker displacement. Standard deviation of residuals from the linear regression line amounted to 0.06 pixels, or about 1:4000 of the viewing field, corresponding to the 90% confidence interval of 1:2500 (commercial brochure data). Very good results are realized with characteristic small markers, which correspond to marker shape of 6×6 pixels.

Marker Identification

In PRIMAS, the principle of marker identification is the same as in the Delft system. This information may be represented on the display, whereby a system is applicable when information is used in real time (feedback applications, clinical testings). At disposal is the WYSIWYG (What You See Is What You Get) type of display. The software package ASYST 3 is used for analysis of measurement data.

In VICON, the system includes a comprehensive program package for 3D calibration, data collection, reduction and sorting (identification of characteristic points) of raw image data, reconstruction of 3D marker locations, numerical differentiation, and graphics. Reconstruction of 3D locations is realized according to the original procedure (1987) called GSI (Geometric Self-Identification). In this application, marker identification in 2D was not necessary—software automatically searches for locations of 3D crossings. Marker identification is operator supervised (sorting) and off-line, but is described as a more precise procedure.

In the example of kinematics measurements by the VICON system, a user-friendly representation is found in the Windows® environment (Figure 4.26). Data are at the user's disposal for further analyses and to conduct the inverse dynamic approach.

Automatic follow-up of markers in time is enabled based on *a priori* knowledge on their application, i.e., anthropomorphic characteristics of their locations, as in the example of ELITE.[15] A special procedure of model definition requires all markers to be connected. Isolated points may also exist. Two attributes are given to markers: priority and weight. Priority takes into account that a certain marker may be masked by another. High priority must be given to the marker which will never disappear, Low priority is given to those which could be occluded during movement. By this mechanism, uncertainty in classification of corresponding markers is resolved. Weights are attributed to the markers. They can be considered to be masses of connections, although they are expressed in round numbers from 1 to 100. They are used for calculation of confidence intervals in predicting the trajectory of a marker during the procedure, where finally, a Taylor series of the trajectory is approximated including third member and derivatives are approximated with finite differences, leading to the following expression:

$$x(n + 1) = \frac{5}{2}\left[x(n) + 2x(n - 1) + \frac{1}{2}x(n - 2)\right] \tag{4.18}$$

Prediction supposes that previous frames were well classified, and so the procedure begins by initialization via interactive graphics (semiautomatic) of the first two pictures and continues by automatic prediction of next marker positions, assigning them to body locations. Overlapping and missing points are solved by means of a modeling structure. For very critical conditions which could cause a wrong classification, a manual editor is available. Borghese and Ferrigno[100] described details of calibration and triangulation procedures in a setup with arbitrary camera positioning.

Other Video Kinematic Measurement Systems

Two additional approaches are described. The first is the Hentschel Random Access Tracking System (HSG 84.330) by Hentschel System-GMBH, Hannover, Germany.[92,101] It uses reflective markers (Scotch® type).

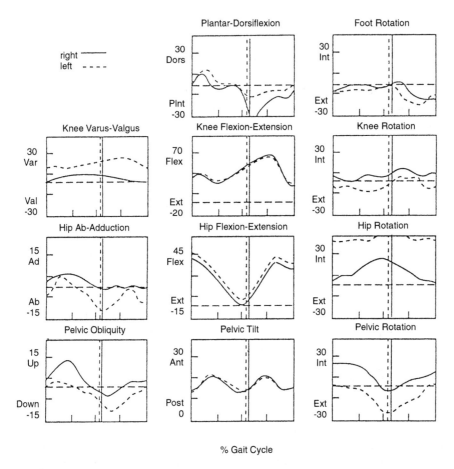

FIGURE 4.26 Human gait kinematics measured by the VICON system. A gait cycle, marked on the abscissa, begins with toe-off. The left column of diagrams shown belongs to the horizontal (coronal) plane, the middle to the sagittal, and the right to the transversal. (From Gage, J.R. 1991. *Gait Analysis in Cerebral Palsy.* Oxford: MacKeith Press. With permission.)

Special cameras are used (by Hamamatsu Corporation, with an image dissector), characterized by the absence of the light storage feature, as in vidicon. There is no need, therefore, to scan a complete scene as is done in vidicon/plumbicon tube cameras to prevent charge accumulation due to ambient lighting. Selective scene tracking is therefore possible, using the Random Access Camera. (A drawback is that strobo-scope lighting is not possible.) When illuminated from the vicinity of the lens, it rec-ognizes light reflectors only. Constituent parts are camera optics and an illumination unit; halogen spotlights are placed around the camera lens. The video interface real-izes input-output to host computer where x, y values are stored.

The basic idea is the following: instead of monitoring the whole image field, as in a normal video mode, only its small parts are monitored. By doing this, a total maximal measurement frequency as high as 15 kHz is achieved. Resultant image sampling frequency is then deduced from the quotient of this total value and the number of points tracked. Video interface of the system brings analog signals from outside to the deflection system which provides deflection in a form of a small rectangle, the size of which is between 0.5 to 4% of the size (linear) of the total image. To each reflective marker, one such small field is attributed, and in one image cycle, one transverses (jumps) in the same sequence from one to the next small rectangle, and x and y coordinates of corresponding reflective markers are sought. By adding the values obtained to the global position of the rectangle in the viewing field, image marker coordinates are obtained. At the same time, this is the position of a small rectangle in the next image.

There are two phases of the procedure. During initialization, the whole viewing field is scanned, the positions of the markers (up to 100 of them) are automatically detected, and their locations are stored in computer memory. Their assignment is done either automatically or by the operator. The second phase is the tracking itself. Control of local scan is with $8 + 8$ bits (for $13 + 13$ bits in total controlling the deflection beam position in a viewing field, equivalent to 8192×8192 pixels). Camera access time for each pixel is below 15 μs.

The device tracks a number of markers with high resolution, with high speed, and in real time. However, according to Furnée's opinion, in spite of the obvious improvement of this system in the sampling frequency of kinematic data, the procedure is sensitive to lost markers. It is primarily used for measuring movement details, not whole body kinematics.

The second approach is using frame grabber-based video kinematics systems, a group of kinematics measurement systems differing fundamentally from those previously described and following the principle similar to that by Winter and colleagues.[95] One solution is the APAS system (Ariel Performance Analysis System), the result of many years of work by Gideon Ariel (in the U.S.), developing computer-supported kinematics measurement procedures with major application in analysis of sports movement structures. The development began in the 1960s in Amherst, Massachusetts, with the high-speed photography method. More recently it has taken place in Trabuco Canyon, California, using video camera-based systems.

Body markers are not required in this method, albeit they can be used if desired. All commercially available video cameras are suitable for application—at least two and nine in all. Prior to video recording, calibration is provided. Recording results in video-taped information. The next step is frame-grabbing: accepting TV images and their digitization one at a time. The frame-grabber has a resolution of 512×480 pixels, with 24 colors and NTSC and S-video inputs. A PAL-compatible unit is optional. Brightness, contrast, saturation, and color can be varied in an effort to achieve a video image superior to the original (it *de facto* represents image processing by noise elimination), but exact specification of a frame being digitized is possible only with the aid of additional synchronization equipment.

Data extraction is done manually, but it may also be done semiautomatically. Manual digitization is a time-consuming procedure (as in high-speed photography). Availability of software for manual digitization is beneficial during the procedure because it always locates approximately the next marker based on the position of previous markers and so work is speeded up, i.e., it is a matter of merely positioning the cursor. Location, brightness, and kinematic parameters, such as velocity and acceleration, are used. In the automatic mode, to locate specific markers, the system uses contrast, color, velocity, and acceleration. Any one color may be used for markers. Different colors may also be selected. The centroid method is employed to calculate the marker center, with a certain tolerance area around the marker. Manual and automatic modes may also be used simultaneously. The system has an affordable price and it is transportable and easily applicable in various situations, but it is also imprecise: the human factor (operator input) is significant in marker location.

Method 2. Systems Based on Position-Sensitive Analog Planar Sensors: Active, coded body markers are used in this method. These are IRLEDs operating in a time-division multiplex. Operation of the method relies on the physical principle of the direct conversion of incoming light, generated by markers, into position-sensitive electrical signals by means of the lateral photoeffect.[29,102–104] The image of an external light source is focused through the camera lens to the surface of a special position-sensitive analog planar detector (PSD, position sensitive detector) located in the image plane of the camera. The detector is rectangular in shape, with bilaterally laid electrical contacts (Figure 4.27). When the detector is electrically connected to the corresponding circuits, each of the generated photocurrents is divided between the two contacts at the detector's edge, and each current is linearly dependent either on the x or y coordinate of the centroid of distribution of an incoming light. By processing the currents obtained, it is possible to compensate for unwanted influences of variable light intensity (the function of distance) as well as for background light (Woltring and Marsolais 1980, according to Reference 29), which both directly influence measurement error.

This is a kind of nonaddressable sensor: one does not need to scan the entire image as in TV methods. The physical principle of the sensor dates back to 1930

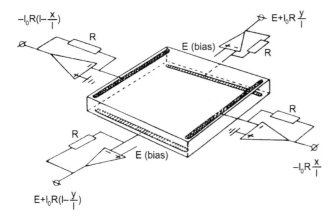

FIGURE 4.27 Analog photodetector of a rectangular shape with linearly laid out side contacts (PSD detector) used in the Swedish SELSPOT system.

(Schottky 1930 and Wallmark 1957, according to Reference 104). Wallmark was the first to discover the possibility of measuring the position of a light spot being tracked at the sensor's surface. Wallmark detectors are among the first optoelectronic position sensors to be applied in optical imaging systems. They had simple point electrodes and therefore high distortion. Bordering the sensors with linear electrodes significantly improved the linearity in determining the location and led to the development of the tetralateral and duolateral types of sensor. Finally, the application of the theorem according to Gear (1969, according to Reference 105) influenced the development of a type of sensor with a circular electrode and excellent linearity for determining position.

A detailed analysis of the physical properties of the 2D diffusion model of a rectangular one- and two-axis lateral photodiode for small signal mode was provided.[28] A nearly linear relationship was determined between the position of the light spot and the corresponding currents; transient responses were discussed as well. Krzystek,[105] citing Reference 28, presented an approximate mathematical development for the duolateral and tetralateral photodiode. If a light source is focused on the sensor, four currents are generated on four electrodes. Theoretically, they are described by the expressions:

$$I_i^x = f_i^x (\bar{x}_p, \bar{y}_p, L, \alpha) \cdot I_o$$
$$i \in [1,2]$$
$$I_i^y = f_i^y (\bar{x}_p, \bar{y}_p, L, \alpha) \cdot I_o$$
$$I_O = I_1^x + I_2^x + I_1^y + I_2^y \qquad (4.19)$$

where the position $(\bar{x}_p, \bar{y}_p)$ of the light spot tracked, the distance L between the electrodes, and the recombination α are parameters of generally nonlinear functions f_i^x and f_i^y, depending on the sensor. The coordinates of the light spot are calculated by means of the expressions:

$$x_p = \frac{L}{2} \cdot \frac{I_1^x - I_2^x}{I_1^x + I_2^x}$$
$$y_p = \frac{L}{2} \cdot \frac{I_1^y - I_2^y}{I_1^y + I_2^y} \qquad (4.20)$$

showing that the position of a tracked light spot may be calculated by applying only four values, but this also depends on the parameters that are a function of the sensor (f_i^x, f_i^y), geometry $(\bar{x}_p, \bar{y}_p, L)$, and physical properties (α). Therefore, the sensor deformation is usually very high and only attains a minimum for duolateral and pincushion-type sensors. Deformation is defined as:

$$dx = x_p - \bar{x}_p$$
$$dy = y_p - \bar{y}_p$$

By applying the principle described, Lindholm (1974, according to Reference 29) first developed a system with more body markers, which represented the beginnings of the SELSPOT device. The method is commercially called SELSPOT (SELective light SPOT recognition) by the Swedish firm Selcom (Selective Electronic Company, Molndal). The method dates from 1974. Progress in technology led to the development of 12-bit resolution per image axis, with 100 μs sampling time per marker. Body markers, IRLEDs, operate in a time-division multiplex. PSD sensors enable the simple realization of measurement systems, working in real time and with high resolution.

However, since the nonaddressable PSD sensor provides intensity-weighted surface integration of all incoming photons, and currents I_1^x and I_1^y correspond to the centroids of the projected light intensities, disturbances of projected light intensity and reflections which may be caused by reflective surfaces result in high image errors, i.e., displacements of coordinates. The SELSPOT system is therefore susceptible to ambient light disturbances. Image geometry is therefore nonprecise and accuracy is limited. The method's precision is higher than 1%.

The light source is controlled to keep continuous light intensity at the detector's surface. This gives a high depth of field without an increase in lens aperture. The system allows the use of up to 120 LEDs, sampling up to 10,000/s, inclusion of up to 16 cameras, and a resolution of 0.025%. Decrease in error caused by reflection is achieved by using optical filters which may limit the detector to IR light, with corresponding LEDs at wavelengths that serve as a source. Ambient lighting may be at normal levels, as opposed to photographic and video kinematic methods which require stronger lighting. The time division multiplexer serially illuminates each of the LEDs, for 50 μs during each data frame. For a 30-LED system, sampling frequency is typically around 100 Hz. Time coincidence between illumination and receipt by camera uniquely identifies each LED. Camera output is digitized typically in 12 bits, so that upper amplitude resolution is 4096, meaning a 0.025% width of field of view. SELSPOT II cameras are used in the Multilab system.[106]

Among the first SELSPOT-based kinematic measurement systems were the ENOCH System, at the Institute of Technology, Uppsala University, Sweden, 1977, and the TRACK (Telemetred Rapid Acquisition of Kinematics) System in what is now the Newman Laboratory for Biomechanics and Human Rehabilitation, Massachusetts Institute of Technology, Cambridge, Massachusetts, (TRACK is commercially available through OsteoKinetics, Newton, Massachusetts.) The ENOCH system was developed primarily for clinical applications in rehabilitation medicine, and it integrated kinematic and ground reaction force measurements and corresponding software support for inverse dynamics.[107] Further development of measurement procedures in the same laboratory included a video camera, but without an inverse dynamic approach. This clinically oriented development is mentioned in Chapter 7.

TRACK was developed successively (the subject of dozens of M.Sc. and Ph.D. theses). The corresponding software support for inverse dynamics is called NEWTON. A standard system for the calibration of the optical camera and transducer was proposed, after which TRACK keeps inherent resolution of the transducer in the image plane in the optoelectronic camera in the viewing field. By using a very broad and precise plotter, an error matrix was developed by comparing physical LED positions in the field of view of the camera with x,y coordinates. The accuracy and resolution of the system achieved are below 1 mm, and below 20 milliradian in angular (rotational) movements, with a working space volume of 2 m³. TRACK has broad application, from scientific research to purely clinical measurements, and in addition to SELSPOT, it also allows inclusion of the WATSMART and ELITE systems.

Using the SELSPOT system, Stokes and colleagues[108,109] developed a method for measuring rotational and translational movements of the pelvis and thorax during treadmill locomotion. The purpose was to identify normal gait kinematics, to be used in the study of locomotor pathologies or gait in developing children. With respect to more conventional measurements of locomotion where movements of extremities— being of larger amplitudes—are usually followed, such measurements are more demanding with regard to accuracy and precision. Two SELSPOT cameras were connected to an HP 1000 minicomputer. Two triangular plates with LEDs at each vertex were fitted to the posterior of the subject's pelvis and thorax via a waist belt and shoulder harness (Figure 4.28). The measurement system also included pressure-sensitive transducers which were fitted to flexible tubing that was firmly attached to the sole of each shoe. The calibration method used a wire pyramidal calibration structure equipped with ten LEDs and followed a defined protocol. A DLT was used for photogrammetry.[84]

Results from eight adults, each performing six trials, indicated characteristic kinematic patterns (Figures 4.29). These results supplement the results of the Berkeley group, clearly showing the functional significance of pelvic movements in gait when three out of six gait determinants referred exactly to pelvic movement: rotation around the vertical axis (the first determinant, see Figure 4.6), rotation around the anterior-posterior axis (second determinant, see Figure 4.7), and translation along the bilateral axis (sixth determinant, see Figure 4.10). Based on measurements of this type, Stokes et al.[109] analyzed the problem of movement control to a greater extent. During the same phases of a gait cycle, the thorax and pelvis rotate in different directions. Some measurement patterns have displayed six changes in rotation/translation, separated by the point of change in direction. Opposite directions of rotation for AX (rotation around the anterior-posterior axis x) and AZ (rotation around the vertical axis z) suggest the possibility of autonomic movement control for

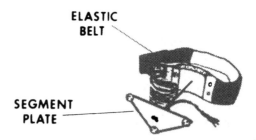

FIGURE 4.28 Plate with actively powered body markers, LEDs, adjusted to the measurement of kinematics of the thorax and pelvis during walking on a treadmill. The SELSPOT system. (From Stokes, V.P. 1984. *Hum. Movement Sci.* 3:77–94. With permission from Elsevier Science.)

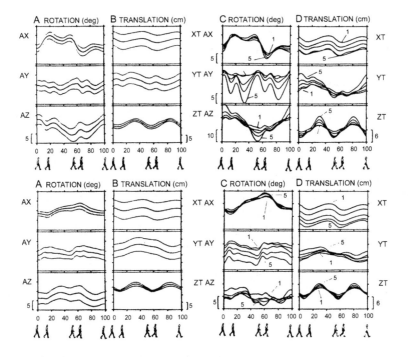

FIGURE 4.29 Kinematic signals of the pelvis and thorax in walking on a treadmill in healthy subjects, measured by the SELSPOT system. Upper diagrams show mean values of pelvic displacement (± 2 S.D.) on a sample of 8 subjects. (A) Rotational and (B) translational displacement, with velocity F. Diagrams (C) and (D) Mean values of rotational and translational displacements, respectively, in a male subject at various speeds. The range of speeds is (1-F $-$ 20%, 5-F $+$ 20%). Lower diagrams, by analogy, show thorax kinematics. (From Stokes, V.P., Andersson, C., and Forssberg, H. 1989. *J. Biomech.* 22(1):43–50. With permission from Elsevier Science.)

the pelvis and thorax. With increasing speed, they adapt differently. The pelvis acts as a gait speed controller (increased stride length with constant frequency). Reciprocal rotational movements of the thorax/pelvis tend to decrease rotational moment of the body and could cause the damping effect to pelvic movement, "smoothing" of gait itself. The constant total peak-to-peak amplitude of pelvic rotation with increasing speed may be in connection with control mechanisms of head stabilization in gait. The SELSPOT method has proven to be noninvasive and precise enough for this application.

The major potential source of error in this method is sensitivity of the lateral photoeffect to light reflection from the environment. Light reflecting from surrounding surfaces, such as from the floor in a room, other body parts, etc., also influences the amount of generated sensor current, increasing it irreversibly, against which measurements are taken as described. Another drawback is the use of active markers that require a portable power supply and cables or a cable connection with a stationary unit (umbilical cord), which is somewhat invasive and also allows the possibility of cross-talk, e.g., with EMG signals. Another drawback is the phase shift between the values of successive measurement channels caused by the time multiplex operation.

Opposed to this, the basic advantage of the method is automatic identification of a number of markers in real time, with good temporospatial resolution (superior to TV), practically 1 kHz, for 5 markers. The system is suitable for real time applications.

Apart from applications in human locomotion measurements, the SELSPOT system is also important for measurement of kinematics in the field of industrial automata and robots, in which application wires such as conductors are not an encumbrance.

Efforts by Northern Digital, Inc. (Waterloo, Ontario, Canada, founded in 1981) are worth mentioning. In 1983, development of a system similar to SELSPOT called WATSMART (WATerloo Spatial Motion Analysis and Retrieval Technique) began, with completion in 1985. As suggested by Woltring,[16] the system took additional reference measurements between LED impulses and calculated coordinates based on the difference between "on" and "off" signals. The method was optimized for fluorescent lighting only. The high accuracy attained was 0.15 to 0.25 mm in 3D in a cube-shaped area with a 0.75-m side.[110,111] Further increase in accuracy seemed unnecessary, however, the system had a significant drawback—the problem of surplus of light, incorrectly termed a reflection problem. Later, 1985 and onward, the firm switched to another principle (see Method 3, the OPTOTRAK system, operative since 1989).

An effort to achieve synchronous detection in PSD-based systems was undertaken at Delft University of Technology, beginning in 1977, with the goal of keeping the inherent advantage of a high sampling frequency attained by these sensors while correcting their drawback of nonsimultaneous sampling.[16] The method uses frequency instead of the time-multiplex for identification of LED markers. LED emissions on the same optical wavelength are intensity-modulated at different frequencies. In this way, distinct LED signal amplitudes are detected which (as in the SELSPOT) correspond to projected marker coordinates. The reflection problem is also eliminated. With regard to sensor noise, the detection principle mentioned acts as a narrow-band filter, and the equivalent noise spectrum width corresponds to the marker sampling frequency. A 240-Hz sampling frequency is inverse to the value of summation time in periodic synchronous detection. Furnee[16] theoretically elaborates the technical characteristics of the procedure in detail with regard to signal processing, modulation applied, noise, and elimination of the reflection problem.

A comparative survey of kinematic measurement systems based on the lateral photoeffect in areas of biomechanics and robotics is given by Krzystek.[105] He accentuated the historical importance of this sensor as being the first to be applied for real time photogrammetry in biomechanics. As opposed to CCD camera-based methods, which are a generation of computer vision systems of a sort that process a large amount of data—therefore they may correctly be classified into artificial intelligence (AI)—position-sensitive analog sensors evaluate image coordinates in a simple way. Therefore, a small computer suffices. The problem was researched extensively by Woltring who attained the speed of 1 kHz in calculating coordinates in real time as early as 1977. Due to the low resolution of a 10- to 12-bit A/D converter and reflections from light sources which are still not eliminated, all these systems and applications in the domain of robot surveying and control attain an absolute accuracy of 0.05 to 1% in optimal conditions. Krzystek's survey encompassed about 15 systems, each with specific characteristics.

Method 3. Systems Based on a 1D Array of Sensors: These systems feature active markers, laid out as IRLEDs. In the COSTEL (Co-)ordinate Spaziali mediante Transduttori Elettronici Lineari) system, manufactured in Italy, LOG.IN sol, markers are IRLEDs which are sequentially lit and housed in a LED unit $10 \times 7 \times 2$ cm in size with a 100-g mass. Up to 20 LEDs can be used, 1 W in power. Optical wavelengths are near the IR part of the spectrum, 800 nm. (A passive marker version and stroboscopic lighting was developed in 1985.) 1D transducers housed in a 3D stereometric unit are used. The sensor element is a linear optoelectronic sensor made of 2048 discrete elements in CCD technology, each 13×12 μm in size, so that the total linear dimension of the sensor is 26.6 mm (Fairchild CCD 143, according to Reference 112). The image of an active LED is projected on the sensor through an anamorphic lens with a toroidal element, while the sensor is in the first focal length of the lens (objective). Lens orientation with respect to the sensor is such that the target marker image is a segment crossing the sensor orthogonally. Therefore, ordinal numbers of enlightened photosensitive elements identify the plane lying along the lens nodal axis and containing the marker. Movements of the marker in this plane do not influence sensor output, so ordinal numbers of enlightened photoelements provide one marker coordinate in a defined referent coordinate system. It is an intrinsically digital output (Figure 4.30). In practice, the number of enlightened photoelements ranges from three to six, depending upon marker position in the field, marker size ($\approx 1 \times 1$ mm), speed of movement, and inevitable image blur. Testing the amplitude distribution of sensor response revealed Gaussian distribution. Since at least three elements are being illuminated, position can be calculated with increased accuracy, therefore resolution attained is 6.5 μm.

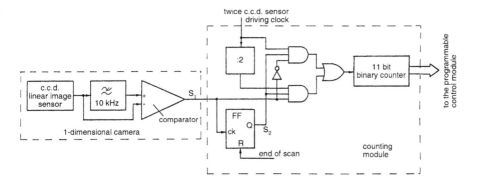

FIGURE 4.30 Block scheme of the electronic circuit supporting a 1D optoelectronic sensor used in the construction of the stereometric kinematic system COSTEL. (From Macellari, V. 1983. *Med. Biol. Eng. Comput.* 21: 311–318. With permission.)

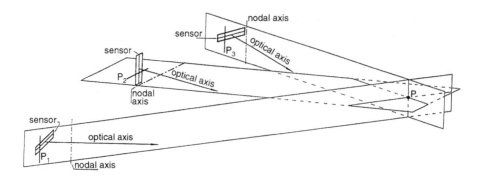

FIGURE 4.31 Combination of three 1D sensors into a 3D unit (system unit) in the COSTEL system. The segments P_1, P_2, and P_3 are the three corresponding images of the point P. (From Macellari, V. 1983. *Med. Biol. Eng. Comput.* 21:311–318. With permission.)

Three 1D sensors are combined to form a SU (system unit), a transduction element for spatial coordinates (Figure 4.31). As opposed to stereophotogrammetric methods described previously, in this method, stereometry is not realized by combining two 2D image planes in corresponding cameras, but three 1D cameras (the linear sensor described previously including electronics) form a stereometric unit. This has implications to trigonometry, so relevant expressions have to be modified accordingly. Bianchi et al.[113] developed corresponding relations and evaluate the accuracy of such a procedure.

3D measurements of human locomotion require 2 × 3 cameras (2 stereometric units). Assuming 8 markers per body side, system electronics enable a maximum of 1000 samples per second (the maximum scanning frequency of a CCD ≈ 9 kHz), but there is a 100 samples per second maximum value in practice due to energy limitation. Calculation of 3D marker location and calculation of velocities and accelerations is done off-line, the time lasting up to 10 s. The real time reconstruction of marker coordinates may be accomplished if optionally a fast processor is added to the control unit. Real time operation at 100 Hz for 6 cameras, with up to 25 active markers, is guaranteed by the manufacturer.

The Canadian firm Northern Digital, Inc. (Waterloo) abandoned the WATS-MART system, and began development and manufacture of a system called Optotrak 3020, aimed to surmount "the reflection problem." IRLED markers (based on a semiconductor chip of surface 1000 μm^2), up to 256 of them, are used, in a sequential strobing regiment enabling automatic marker identification in real time. Maximum sampling is 3500/s.

An addressable linear 1D optoelectronic sensor with a 2048-element CCD with a custom-designed anamorphic lens is used. Sensor accuracy is ± 1:50,000, and resolution is better than ± 1:20,000. Since the lens system does not require focusing, the lens elements and linear CCD array are mounted in a compact unit. A standard device outline includes three such sensors, paired with three lens cells, firmly mounted in a mechanical unit: a 1.1-m bar (Figure 4.32). Inside each of the three lens units, light from an IR source is directed to the CCD sensor and measured. All three measurements determine, in real time, 3D marker location. The unit can be spatially oriented

FIGURE 4.32   Optoelectronic stereometric kinematic measurement system (OPTOTRAK).

as desired. A practical feature is that system calibration is done by the construction itself. The device is stable for standardized measurements in fixed conditions. Another standard device layout includes a 2D sensor, which may also be desirable, since such units may be combined for measurements in a larger space. To keep up with the high speed of data acquisition, computer support is based on a RISC technology.

RMS accuracy of 0.1 mm is achieved, with 0.01-mm resolution. Working space is from 1.3 × 1.3 m, at a 2.25 m distance, to 3.5 × 2.5 m, at a 6 m distance. Accuracies of 30 μm in space of 1 m³ are achieved (2D camera). Each camera is able to locate a point from 1.8-m distance with 1:64,000 resolution and 1:20,000 accuracy across the field of view. Besides human kinematic measurements, the instrument is commonly applicable to engineering problems of tracking and control in automation systems.

Bianchi et al.[113] presented arguments in favor of 1D DCD sensors with respect to 2D, such as used in video methods. In the first place, it is certainly more demanding from a numerical aspect to detect a light spot on a square matrix of, for instance, 256 × 256 = 65,536 photosensitive locations, than on a linear matrix with only 2048 photosensitive locations. This limits overall frame rate significantly in the first class of systems. Furthermore, attempts were made to improve spatial resolution achieved in matrix sensor-based systems by centroid calculation procedures, but resolution is basically superior in 1D sensors. The spatial resolution achieved is 0.025% of the measurement field, precision is 0.02%, and accuracy (maximum absolute error) is 0.1%. The systems function well in conditions of lighting disturbances and reflections. However, 1D sensor systems require a more demanding calibration procedure. There is also a distortion error present, caused by the anamorphic lens.

The following summary is appropriate at the end of a presentation of optoelectronic methods (video and others). In principle, automatic digitization of image coordinates is provided in these systems. The exception is frame-grabber type systems. In some systems, automatic marker identification is carried out, and this is partially possible in frame-grabber type systems as well. This is of great practical importance. In general, body markers may either be coded or non-coded. They may be either active (power supplied) or passive. One subgroup of optoelectronic methods may be called TV methods—systems using a standard TV camera connected by electrical means to a computer through a corresponding interface. Considering the nature of the TV image, it is essential for TV methods that the camera measures 2D image coordinates of all markers simultaneously, more precisely with corresponding delays, picture by picture, besides all the drawbacks and advantages of this procedure. A number of researchers share the opinion that this group of methods is the future of human kinematics measurement.

4.2.1.4 Light Scanning-Based Methods

Contrary to stereophotogrammetric methods, in a sense, light scanning-based methods accomplish an active detection. In these methods, measurement sensors do not receive light from the environment passively, as in methods previously described, but signals are sent toward a measurement object, and its response is in some way detected. By actively covering a working area using three units, 3D measurement can be realized. Therefore, direct 3D reconstruction is achieved, without the need to apply the analytical photogrammetry procedure. Markers in systems of this type may be passive, reflective, or optical detectors, which determines the technical layout of these systems.

The CODA-3 (Cartesian Optoelectronic Digital Anthropometer), manufactured by Charnwood Dynamics Ltd. from Loughborough, Leicestershire, Great Britain, was initially designed in 1970. During the first decade, work took place in the department of Ergonomics and Cybernetics at Loughborough University. A 1973 prototype used custom-designed compound cylindrical lenses in three cameras rigidly mounted on a fixed stereo base.[114] The cameras also had a custom-built hybrid A/D photodetector array pattern in the focal plane, with which the image of the LED marker was encoded. The aim of the project was to design a 1:4000 resolution system. However, this took place before the advent of linear CCD arrays, and a resolution of only 1 part in 1000 was achieved.

The firm then moved to the mirror scanning technique, capable of angular resolution in excess of 1:50,000. Since 1980, development continued in a commercial environment, and eight-marker systems were delivered in 1983. These used the same fixed stereo-base as in the early prototype, but with mirror scanners laid out as rotating multifaceted polygonal mirrors instead of cameras and passive corner cube prism reflectors as markers. A color coding approach was used in which each prism marker was fitted with a unique ocular filter.

The device marketed as CODA MPX 30 contains three scanning mirror assemblies rigidly mounted to a fixed baseline within the scanner unit (Figure 4.33). The markers take the form of pyramidal corner cube prisms, four mounted back to back and equipped with a unique color filter. Two scanners are positioned 1 m apart at the

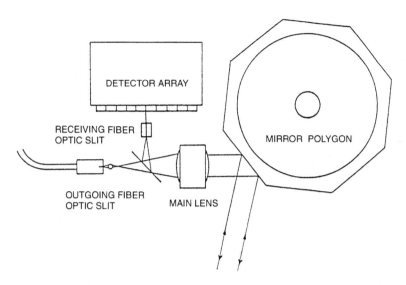

FIGURE 4.33 CODA scanner, including three mirror units and optical arrangement within each scanner. (From Michelson, D. 1990. *Image-Based Motion Measurement,* J.S. Walton, Ed., SPIE Vol. 1356:26–37. With permission of the International Society of Optical Engineers.)

sides of the chassis of a compact unit. By rotating around the vertical axis, they send out synchronously thin vertically oriented sheets of light, 1 cm in width, which sweep horizontally across the field of view. These provide signals which are used to triangulate for marker distance and transverse horizontal coordinate. The middle scanner, mounted midway down the scanner, rotates around a horizontal axis and sweeps a horizontal sheet vertically, thus providing signals from which vertical marker position is derived. Each mirror scans a 40° angle, at a rate of 300 scans/s.

When a beam crosses a marker, a light pulse is reflected, returns along the same path, and is detected by photodiodes positioned in the scanner unit. The phase of the return pulse, with respect to mirror rotation, is a measure of marker position in the angular field of view. The time of each scan is quantized by a counter into 40,000 increments and the time of each returning pulse is thus a measure of the angular position of the marker in the scanner field of view. Scanner output signals are fed into a computer where 3D marker coordinates in each scanning period of 1.67 ms are calculated in real time. The system is free of parallax error. The device's hardware incorporates seven microprocessors which use this timing information to calculate the global spatial coordinates (X,Y, and Z) of each marker in real time. The time lapse from scan to calculation of output is 10 ms.

The key to high spatial resolution in the CODA-3 is that position measurement is converted by the scanning process into a time measurement which can be carried out with great precision. The master clock of the system is phase locked to the signal from a rotary grating rigidly mounted to the rotating mirrors, so measurement has zero drift. Also, by using mirrors to scan the field of view, the light path through the other elements of the optical system remains stationary and in the optical axis whatever the position of the prism marker may be. Consequently, CODA-3 is free of the distortion error normally present in systems which depend on lenses to form an image of the whole field of view. The instrument is factory-calibrated.

Each marker is uniquely identified by a color filter which is recognized by an optoelectronic color decoding system, using splitting optics and diffractive gratings. In this way, marker identification is automatically established after start-up or if the marker is temporarily obscured during measurement. If a marker is obscured, its coordinate outputs retain the last-measured value until it reappears.

The basic resolution of each scanner is 1:16,000 across its scanning field. For a given distance Z from each scanner, the width of its field is 0.8 Z. Three-dimensional measurement is defined in a region where scanners A and C overlap. It can be seen that the effective width W of the measurement field is given as W = (0.8 Z − 1) m. Resolution in X and Y directions remains close to the basic resolution for each scanner. Resolution in the Z direction is divided by the ratio of distance from the base line. The basic resolution figure refers to individual data samples being taken at 600 Hz. Lower effective output rates can be selected in submultiples of 2. In this case, the device uses a high 600-Hz rate, so that it averages across a corresponding number of "raw" data samples to achieve the selected lower rate. This improves the final resolution in proportion to the square root of the number samples over which averaging occurs. Up to 20 markers can be tracked with combination of color identification (12 colors) and an intelligent software post-processing procedure.[114]

3D global marker coordinates are obtained in real time which is an advantage compared to the majority of other systems (system calibration being inherent to mirror configuration). The advantage of the method is that it is completely automatic, with passive markers.

A disadvantage of the method is that data are acquired in nonequal time intervals, depending on the target monitored, and consequently a time skew appears, necessitating additional data processing. Also, the outline of the device depends on fine mechanics and not on microelectronics, so it does not have as bright a perspective as consumer electronic products.

Besides these methods, there are a number of additional stereometric methods commercially available. Chao[78] described an ultrasonic stereometric system. Ultrasonic transducers are attached to body segments, producing ultrasound pulses, and they serve as three non-collinear reference points. Three linear, orthogonally oriented microphone sensors form X,Y, and Z global coordinate axes and uniquely determine spatial location of transducer (Figure 4.34). Photogrammetry is not needed in this method and there are no lens errors. Spatial resolution achieved in the working space of 1 m³ is around 0.1 mm. Maximum sampling frequency is 140 points/s. If two or more transducers in the same area are used, time multiplex is needed. Besides a small working area, the system's drawback is sensitivity to metal-induced noise. Additional ultrasound-based systems are marketed as well.[115]

Diverse physical principles in addition to ultrasound may be employed such as those based on magnetic fields,[116] etc., as well as various signal processing methods. However, in the field of biomechanics, those stereometric principles described previously are the ones mostly used.

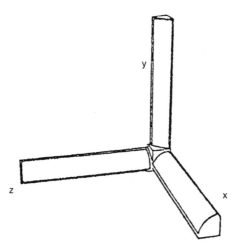

FIGURE 4.34 Ultrasonic digitizer used for locating points in space. (From Chao, E.V. 1978. *CRC Handbook of Engineering in Medicine and Biology. Section B: Instruments and Measurements*, Vol. 1, B.N. Feinberg and D.G. Fleming, Eds., Boca Raton, FL: CRC Press, 385–411. With permission.)

4.2.1.5 Stereometric Methods—Final Considerations, Signal Processing Aspects, and Computer Vision Issues

Stereometric kinematics measurement methods of human movement are a subject of constant, dynamic development. Their crucial features shall be summarized and compared, and their relation to the field of artificial computer vision shall be identified.

There are some comprehensive survey articles which give a critical comparison and evaluation of kinematic measurement systems. Stussi and Müller[117] provided a systematic comparison of the features of eight commercially available kinematics measurement systems at that time, according to the following parameters: type of sensor, type of marker and lighting, angle of camera and field of view, mode of automatic identification and tracking of body markers, calibration procedure, sampling frequency, measurement precision, accuracy, linearity, available software, and price. They suggested one type of quantitative validation of the overall performances of the respective systems. Furnée[92] also compared commercial systems by analyzing technical aspects of data capture, signal processing, and performance separately with respect to conditions of practical application in the presence of noise. He particularly stressed the problems of subject-measurement system interface with regard to markers: practical layout, possible obstruction during measurement, etc. Koff,[118] based on an understanding of the technical characteristics of about 20 commercial systems, described features and critical moments of the technical layout of the image detection procedure via microelectronic sensors and developed general principles for the realization of the photogrammetry procedure. He also discussed solutions for 2D/3D data conversion and the spatial calibration of the entire kinematic measurement setup, also suggesting corresponding tests for this purpose.

At the beginning of the 1990s, the most popular systems were the SELSPOT II system (about 150 users, including the field of robotics) and VICON (about 100 users). Next came the ELITE, OPTOTRAK, and CODA systems. The very popular high-speed photography method was mentioned by Furnée only historically, when discussing Marey. In a certain sense, however, the approach has not changed significantly since Marey's time because in present-day systems the problem is also reduced to the measurement of spatial locations of a limited number of markers, which are in effect an abstraction of the moving skeletal structure. Data are reduced and the moving body is analyzed by a description of marker locations, not via shape recognition of particular segments. This kind of classical biomechnanical approach contrasts with the biostereometry approach which is interested in a detailed measurement of morphology, i.e., the shape of the quasi-stationary surface of the human body.

With regard to the field of view, camera-based systems offer the most suitable conditions due to the possibility of choice of objectives adjusted to the situation (VICON, ELITE, PRIMAS, Henschel). The disadvantage of these systems is that it is necessary to repeatedly provide linearization for each combination of objective-blend. SELSPOT and OPTOTRAK have a camera angle of 25°, i.e., 30°, and distances from 0.5 to 40 m (depending on the number of LEDs), i.e., 2 to 8 m, respectively. CODA, however, has the smallest field of view: 30° to 35° and a distance of 3 to 8 m.

Along with the rigid segment paradigm adopted in biomechanical modeling of the human body, a practical solution has been put into practice with respect to the application of body markers. Marker clusters are fixed to a low inertia basis,[38] hence the problem of individual markers moving due to skin movement is circumvented[119] and definition of the 3D location of segments in space is possible. This is undoubtly much more suitable than invasive methods such as the one using pins transcutaneously inserted into the bones of a subject described by Saunders et al. However, the (non)invasiveness of such modern procedures may also be questioned. Furnée[94] underlines the difference between electrogoniometers and optoelectronic systems with marker clusters. What makes the later noninvasive is the absence of external connections (mechanical leads). Whether the power needed for active markers (as opposed to passive) is supplied by cable or not is not a serious problem.

As discussed in the descriptions of particular methods, many kinds of markers are used. Active markers, typically LEDs, have the advantage of enabling time multiplex, and as a result unique identification, but they require cable connections to a power supply. Passive reflective markers do not require connections which is an advantage, but they require "intelligent" recognition procedures. Passive marker illumination is provided either by visible or by IR light, mainly using a strobe. Markers are retroreflective — the average time resolution is limited to 100/s, with satisfactory spatial resolution and accuracy.

With regard to the outline for the sensor procedure of stereometric measurement, the following may be concluded. Within the subgroup of stereometric methods, the dilemma is whether to opt for analog or digital (linear and matrix) sensors. A comparison of analog and digital sensors favors digital. Researchers at the Zagreb Institute "Ruder Bošković" came to a similar conclusion when developing compatible applications of optoelectronic sensors in noncontact measurements of surfaces, i.e., as opposed to locomotion, in quasistatic situations, which required high spatial resolution.[120-122] In CCD-sensor systems, the marker detector in automatic systems is usually implemented by hardware and may be based on principles ranging from simple threshold detection to procedures of pattern recognition or quadrature edge detection. Koff[118] also mentioned the application of the Sobel operator in this area. With the introduction of CCD, technology (layouts up to 600×800 pixels were realized until 1991) increased accuracy tenfold. The largest increase in accuracy for determining marker coordinates was facilitated by the application of centroid calculation and cross-correlation. The trend is toward high-resolution CCD cameras. In this area, the principal question posed is "does one have to increase accuracy and precision of stereometric systems to perfection since there are other error sources apart from mere measurement that contribute significantly to errors in the total realization of the inverse dynamic approach?"

As far as signal processing is concerned, stereometric kinematic measurement methods may be considered systems of quantized amplitude and sampled data. Body marker coordinates are data that are subjected to the process mentioned. In principle, this does not differ from the classical A/D conversion procedure (Section 3.2); it consists of a series of A/D converters. It is therefore useful to identify and quantify error sources in a procedure by using the usual criteria as is done in standard industrial A/D converters. These are those in connection with amplitude effects, such as noise contamination, resolution, and accuracy (often expressed through linearity) and those in connection with time sampling skew, jitter, and aperture time. Concerning phase shift due to sampling, only a solution with TV systems with a strobe is characterized by simultaneous marker sampling which is an advantage. (As discussed earlier, the same is realized at the expense of a highly intelligent recognition procedure, classification, and marker tracking and also includes the problem of dropped samples.) Because the goal of all stereometric systems is to observe more markers, an analogy with the array of A/D converters or a group of simultaneous or time skewed sample/hold circuits, multiplexed to one A/D converter, is imposed. When markers are not sampled simultaneously (time multiplex, as in the methods based on analog planar sensors and especially in the methods with a changing time interval, such as in light scanning-based methods), problems appear. Correction software may partially solve these problems.

As a rule, the sensor procedure is complex and even contradictory in a sense. There is a tendency to bring as much information as possible from the hardware into the computer and to subject it to processing for the detection to be performed, since it is then possible to estimate marker centers more accurately, but, this advantage contradicts the idea of data reduction. Fast development of hardware layout and decrease in price of processors and memory will more than compensate this shortcoming in the future. The trend is for stereometric systems to develop in the direction of image processing, where super-fast frame-grabbers are applied.

Concerning marker identification and tracking, some systems (VICON, ELITE) offer software support for non-self-identifying markers. The idea of automatic marker tracking by using neural networks for assigning markers on the basis of files of previously tracked data has been proposed.[123]

Description of the A/D conversion procedure in Chapter 3 also includes definitions of time and amplitude resolutions, as well as noise (Equations 3.7 to 3.10). Two
vital papers by Lanshammar[124,125] address the problem of differentiation of kinematic
data (Section 4.4.2). Assuming a signal with a limited frequency spectrum and additive white noise, signal derivatives calculation, which will be necessary later, has to
be taken into account early during sampling which suggests increased sampling frequency. This decreases noise which is important for differentiation which follows. In
connection with this, Furnée suggested the following quantitative validity criterion
for the sampling system of kinematic data contaminated by noise. The spatiotemporal resolution of a measurement system Q is defined as:

$$Q = \sqrt{f_u}/p \qquad (4.21)$$

f_u denotes the sampling frequency and p denotes precision. According to this criterion, the best stereometric system is PRIMAS, where the Q value is 13,000.

Pedotti and Ferrigno[123] discussed the problem of a stereometric kinematic measurement system functioning in real time. The quantity and speed of incoming data
(data throughput) are defined by the product of the number of markers used, the number of sensors, and sampling frequency (f_u). In general, dedicated DMA protocols can
be used, GPIB (IEEE-488) and Ethernet protocol. Real time data storage is proportional to the product of the product mentioned and the duration time for the acquisition process of kinematic data.

In TV camera-based methods, increasing automatization of stereometric procedures and development of the level of intelligence of measurement systems indicates
a trend concordant to the area of artificial computer vision.[126] In this field which grew
as part of the industrial robotics, automation, space research, and military industries,
there are three main research areas: image processing, pattern recognition, and image
understanding.

Image processing is concerned with the transmission, storage, enhancement, and
restoration of images. A number of techniques are available in which feature detection is realized, needed later for recognition and analysis, i.e., smoothing, gradient
detection (being mutually inverse operations), and thresholding which are all localized techniques (with respect to the position in an image). Thresholding is usually
done first by means of gradient detection and then by the segmentation process.
Segmentation can be, in principle, done either by edge finding or region growing. In
the practical realization of such procedures, the presence of VLSI technology
enabling realization of parallel structures is very important. In this book, from the
aspect of application, the procedures of thresholding, such as are applied in edge
determination in some commercial systems, are of primary relevance.

Pattern recognition is concerned with image interpretation within the context of
previously defined marker classification. The fact that objects which generate the
image are usually 2D and are either formally defined or marked as variations to a prototype is important. Four basic techniques can be applied: template matching (as
applied in ELITE), feature analysis, relational analysis, and syntactic methods.

Image understanding addresses the perception and interpretation of 3D scenes.
In the field of human locomotion measurement area, these issues are defined by the

biomechanical methodology for movement analysis itself (Section 2.1). An advanced example, covering 3D kinematics and also the kinetic function of locomotor apparatus (by muscle action), is the commercial system MusculoGraphics. In industrial robotics, representation of a 3D scene may either be model-driven or picture-driven.[127] A direction can be envisaged where kinematic measurement systems will evolve into completely marker-free methods. Possibly this would solve the dichotomy between the field of human locomotion measurement and static biostereometry: it could be called dynamic biostereometry. One possible development in this direction is presented in Section 7.2.

To conclude this presentation of stereometric methods in measuring human movement, the words of caution by Atha[74] seem justified. He pointed to the danger of non-critical researchers applying automated systems. The ideal critical user should be a qualified professional in fields such as optics, dynamics, and photogrammetry,[118] but also in electronics and anatomy. An analogous situation presents itself in the field of computer application of statistics in processing experimental data of various origins. Non-critical users often rely on "the computer will process the data," in spite of the fact that perhaps they do not fully understand the essence of the statistical processing methods used.

4.3. ACCELEROMETRY

The third group of kinematic measurement methods is accelerometry. As opposed to electrogoniometers and stereometric methods, the devices in this group do not provide measurement of displacement, but of the spatial acceleration of a rigid body. In this manner, 3D measurement of human kinematics is made possible, in principle. According to Euler's theorem, the general displacement of a rigid body with one fixed point can be obtained by a rotation around a properly oriented axis passing through the fixed point. The direction of the axis can be determined by two angles, and so the addition of a third rotation around the axis completely defines the body's motion with one fixed point. This information can be supplied by three independent measurements of orthogonal accelerations. For additional motion of the body-fixed point, three additional measurements are needed, with respect to the inertial reference system. Therefore, in general, the spatial movement of a segment may be determined by means of six appropriately directed linear accelerometers (Morris 1973, according to References 78 and 128).

The sensor employed in this method is a linear accelerometer, characterized by an output signal proportional to the acceleration of the sensor in one direction. Using a cantilever beam, a small mass is attached to the moving body whose acceleration is to be measured. A set of strain gages attached to the beam is used to transduce beam deflection into an electrical signal. As the body accelerates, the beam is deflected and the amount of deflection is related to the acceleration of the mass. The sensors are piezoresistive transducers, made of silicon glass, allowing miniaturization of the sensor to approximately 1 cm. Sensors are positioned to react to deflection in a certain direction and are connected in Wheatstone circuits. Three sensors, whose sensitive axes are mutually perpendicular, combine to form a triaxial accelerometer. Solutions are designed in the form of a silicon chip.[82] The frequency band of such sensors is from 0 to 1000 Hz. The accelerometer output is the vector sum of kinematic acceleration and the acceleration due to gravity.

In principle, by properly positioning accelerometers to selected points on the subject's body, spatial accelerations of these points as a function of time may be measured. These data could then be fed to the computer via an A/D converter. If initial conditions are known, velocities and accelerations of monitored points can be calculated through numerical integration procedures. Numerical integration procedures are significantly less critical with respect to noise because integration *de facto* attenuates noise, making it a great comparative advantage over methods which, while measuring displacement, require the application of numerical differentiation of signals (see Section 4.4.2).

There are a number of points essential to the realization of this procedure. Prior to integration, raw accelerometer signals have to be filtered — electronically or digitally. The choice of filter cut-off frequency depends upon the characteristic frequencies of joint movement. In cyclic movements, error caused by signal drift may, in practice, be corrected by equaling signal values at the beginning and at the end of the cycle. The resulting measurement of acceleration along each accelerometer axis can be obtained by subtracting gravitational effect. If one chooses the origin at a fixed location on a moving segment, and calculates translational movement for the origin, the complete movement of a rigid segment may be defined. This procedure of data reduction refers to the movement of one segment only. To determine relative movement in joints, the procedure must be carried out for other connecting segments too. This approach is for an ideal case which, however, is not practical in realization, and there are considerable problems with initial conditions. Therefore, there are a relatively small number of papers which report on the use of accelerometry to estimate forces and moments in joints during locomotion, especially for the body as a whole.

Ladin and Wu[82,129] described an approach aimed at increasing the accuracy of the procedure for estimating forces and moments which combines measurement of position and acceleration. An integrated kinematic sensor was used which possesses markers for position measurement, monitored by a WATSMART system (Section 4.2.1), and a triaxial accelerometer for direct measurement of segmental accelerations. The sensor position and orientation in laboratory coordinates (determined by TRACK) were used to determine instantaneous orientation of accelerometers in a gravitational field. This information was used to subtract the effect of gravity from the accelerometer output signal, leaving only kinematic acceleration (second derivative of position vector). This value, when substituted in dynamic equations describing the movement of a rigid body (Section 2.1), gives force estimates in joints. The integrated kinematic sensor was designed in order to improve the quality of estimation of joint forces based on the classical inverse dynamic approach. The authors recommended the use of a combined method of this kind in all cases where significant impact forces are manifested.

In view of the points mentioned, application of the accelerometric method to human locomotion measurements is limited. In conclusion to the description of kinematic measurement systems, recent recommendations by the International Society of Biomechanics (ISB) for human kinematics measurement from the *Journal of Biomechanics* are given in Appendix 1.

4.4. KINEMATIC DATA PROCESSING

"Raw" kinematic data, signifying displacements of body landmark points (and all methods except accelerometry measure displacements), do not carry large information content, but are only a description of movement of sorts. Some of the kinematic quantities, however, may be of certain value if measured under standard conditions of performance of the locomotion concerned and may help the expert, physician, or coach to evaluate performance. Previously mentioned phase diagrams showing the dependence of two kinematic variables, such as joint angles, are an example. Furthermore, the mere visual inspection of graphical representations of some kinematic data (curves) can, for an experienced kinesiologist, afford insight into the technique of sporting movement that he does not have by merely viewing the movement structure. In this way, a comparison of top athletes is possible with kinematics since recordings may be made during competitions and are completely non-invasive. This is a continuation of events in the history of locomotion measurement from Leland Stanford's and Eadweard Muybridge's time. In general, however, the true potential of kinematic measurement data lies in their application possibilities, via an inverse dynamic approach, in the mathematical estimation of kinetic movement quantities (Section 2.1). The goal of processing is to calculate accelerations and to combine these data with estimated inertial body features (Expressions 2.1 to 2.10).

Accelerometers provide acceleration values directly. There is no doubt that this is their advantage. However, as discussed earlier, a global approach to body kinematics using accelerometer measurements (about 30 points) is technically impractical. A systematic solution to the measurement of global body kinematics is therefore only provided by methods with 3D displacement quantities as their output, i.e., stereometric methods. The basic processing operation to which kinematic signals have to be subjected is, therefore, derivation, which, in practice, is numerical differentiation. This procedure must be applied to measurement data twice, consecutively, according to Expressions 2.1 to 2.10. Inertial forces and gravitational force in such a procedure (which also holds true for moments) practically contribute to fast sporting movements and jerky (third derivative) pathological movements, since there are large changes in velocities and accelerations. In slow movements, inertial effects may be neglected, i.e., one may assume a quasi-static approach. The issues of kinematic data processing are more critical for the first movement categories.[130]

A classic treatment of kinematic signals includes the following series of procedures:[30]

1. Reconstruction of locations of observed body markers in time (which is included in 3D kinematic measurement itself)
2. Signal smoothing/interpolation
3. Application of the human body model of the inertial segment type, through which anatomical signals are "produced" from measured signals
4. Differentiation of these signals

Due to the inevitable presence of noise in "raw" kinematic measurement data, and because of its characteristics, the direct application of the numerical differentiation procedure, twice in succession, will result in an increase in noise components. Namely, numerical differentiation has a high-pass filtering effect, i.e., a relative increase in higher frequency components of the total signal. Suppose there is a sinusoidal signal of frequency f with the superimposed component of (imagined) measurement error represented by the sinusoidal signal of frequency 10xf and 1/10 of the amplitude.[131] Accordingly, signal/noise ratio amounts to 10:1. After the differentiation procedure has been completed, the amplitude of (imagined) noise would become equal to the amplitude of the signal, while it would become even 100 times larger than the signal amplitude after the second differentiation, i.e., the amplitude of the first derivative of sinusoidal signal is proportional to f and the second is proportional to f^2. If this ideal situation is transferred to human movement kinematic signal situation, we see that the problem is that the frequency spectrum of the useful signal is generally of low frequencies, while noise is characterized by a white spectrum, i.e., with equally distributed power over the entire spectrum. It is therefore evident that such a situation may introduce noticeable disturbance in the estimated values of acceleration and may even completely mask the useful information. It is therefore of interest to develop and perfect mathematical procedures and techniques for signal processing which would minimize such unwanted errors.

Prior to the differentiation procedure, kinematic signals have to be—among other things—low-pass filtered in the aim of eliminating noise. The low-pass filtering procedure should be a compromise between noise elimination and keeping useful information in the signal. In more traditional technical areas, e.g., such as radiocommunications or automatic control, the frequency spectra of the useful signal and of noise are known *a priori*. In human movement kinematic signals, this, however, is only true up to a point. Therefore, the classical procedure of optimal separation of noise from the signal via the method of least squares using a Wiener filter cannot be applied.[107,130,132] The numerical procedures which were applied most in this area, using high-speed photography to begin with, and other kinematic measurement methods in use today, taking into account real knowledge of the features of the locomotor process and real measurement conditions (the problem of omitted samples, for instance), will be presented (Section 4.4.2.). According to what has been stated, numerical differentiation might, in this case, be considered an "ill-posed" problem (according to Woltring), since its realization on a signal with an unknown frequency spectrum is required.

4.4.1 SOURCES OF ERRORS IN KINEMATIC MEASUREMENTS

From the standpoint of measurement theory, errors may, in general, be classified either as systematic or random. Systematic errors in measured kinematic quantities of human movement, i.e., errors introducing a consistent bias in data, are usually characterized by frequencies below 10 Hz and therefore, cannot be eliminated by a smoothing procedure since this would interfere with the useful signal. Random measurement errors are customarily presumed to follow Gaussian distribution (zero mean value and a certain value of variance), and according to the "central limit" theorem, such noise is additive and not a function of the signal. This section lists the sources of errors of particular kinematic measurement methods. (Their more detailed description, i.e., quantification, is beyond the scope of this book.)

Errors in Electrogoniometry: Basic errors in this measurement method result from non-ideal tracking of the joint axis(es) during the measurement procedure and also from movements and vibrations of the device during measurement.[75]

Errors in Stereometric Methods: This group of methods has certain errors in common. As low range photogrammetry is mentioned here, c (camera focal length) is a function of the object-camera distance, according to the lens formula connecting the distance between object and image with the focal length (Expression 4.2). For a fixed camera, changes in distance cause image blur and distortion. Blur influences image details in an unpredictable way (also positional accuracy). Distortion causes systematic image deformations which can be taken into account by the calibration procedure. Image sharpness will, in practice, be satisfactory.

Projective Equations 4.3 and 4.4 formalize that P, p, and C are on a common straight line. Because the camera (lens) is not ideal, these expressions may be used only after correction of "raw" measurement data. Lens thickness results in two different projective centers, C and c, for each of the two coordinate systems, respectively. Angular relations are preserved. Thus, only an extra translation and possibly rotation are involved. The projective relations in fact involve directions rather than absolute points in two coordinate systems. In Section 4.2.1, the mathematical model of the photogrammetric procedure shown is valid for ideal projective relations (I order theory). The model of deviation from the ideal case is called a II order theory, by which one has to estimate unknown parameters, in some optimal sense. Woltring[85] analyzes the mutual connection of errors in such a system in the function of the layout of the control grid. The spatial grid is a certain concrete 3D structure, corresponding to what has been mentioned in previously described methods. Resulting mathematical expressions are usually nonlinear and not solvable explicitly, as in calibration, and also during measurement. Iterative and implicit estimation techniques are used.

In many photogrammetric applications, for instance, aerial terrain surveying, system calibration, and target measurement may, even must, be conducted simultaneously. On the contrary, in real-time low-range photogrammetry, it is possible to separate calibration and measurement. The procedure begins with camera calibration by means of a calibration grid with a certain number of control points which have exactly determined spatial locations. The number of control points, as mentioned earlier, is at least 12, but it is better to have a larger number. Mechanical accuracy of the layout of the calibration grid determines precision to about 1 mm (Hatze). Once the calibration grid is recorded, those cameras cannot be moved further. Then the studied movement structure is recorded. Numerical procedures of calibration and measurement are iterative.

Lens errors are, at most, radial, and additionally there may also be tangential errors, normal to the radial direction.

Camera calibration and the reconstruction of target location (i.e., space intersection) are treated in a dualistic or simultaneous way. Direct linear transformation

(DLT)[84] represents the most popular approach today, accommodating all forms of image distortion and enabling random camera positioning with regard to a global coordinate system. DLT requires a complex, 3D calibration grid with known control points. Marzan and Karara (1975, according to Reference 83) have generalized DLT to accommodate the model of non-linear distortion. There are a number of excellent 3D techniques published, with error in the range of 0.01 to 5 mm. Within this context it is worth mentioning that there has been noticeable increase in the use of the DLT procedure, especially recently. Search of the SCI data base (September 1997) showed that the pioneering article by Abdel-Aziz and Karara[84] had been cited slightly more than 100 times, mostly in SCI and significantly less in SSCI referent base. This can be explained by the fact that significantly higher kinematic measurements as a method belong to the natural, technical, and medical sciences, in comparison to social sciences. From this number, approximately half of them appeared in the 5 years prior to 1997, and the trend was highest in the first half of 1997 (10 citations). This leads to the observation that the algorithm is being applied extensively in the world, as also is the practical use of photogrammetric calculation in the realm of kinematic measurements.

Considerations so far have assumed fixed camera positions. The possibility of moving cameras during measurement and so additionally broadening active space may be very useful. In this case, the mathematical model in the photogrammetric procedure shown needs to be modified. Allard and colleagues[133] presented one such solution: camera movements in two directions are foreseen, introducing a novel variable into the mathematical description. In general, the positioning of the camera in this kind of measurement will depend on concrete application. In principle, small parallax angles have to be avoided, to not decrease the depth accuracy below a permitted limit.

In the high-speed photography method, there are additional sources of error: spatial misalignment of the camera; parallax error due to the object being outside the photographic plane; stretching of film or imperfect registration of film in the camera; movement of the camera or projector; distortion due to the optical system of camera and projector; graininess of the film; error in recording of time or spatial measure; finite precision of the digitization procedure; and error of operator estimation and parallax error in locating joint rotation axes.

Relevant to all kinematic methods are errors due to marker movements relative to the skin, analyzed in detail by Lamoureux.[119] Furthermore, marker movements may occur with respect to the joint rotation axis as well (either due to skin movement or axial rotation of the segment). Woltring[31] considers this part of photogrammetry which deals with error propagation more complicated than mere estimation of parameters, i.e., the system's state.

Errors in Accellerometry: Basic error sources are invasiveness due to the accelerometer's mass, drift, and initial conditions.

All sources of error in the kinematic measurement methods mentioned here relate directly to the procedures of actual measurements and are the result of their non-ideality, finite (real) accuracy, and precision. The next section will describe numerical differentiation procedures of measured kinematic quantities. Additional errors in the inverse dynamic approach that result from the choice of biomechanical body model (simplified geometrical shape and segmental inertial properties) are not within the scope of this book.

4.4.2 FILTERING AND NUMERICAL DIFFERENTIATION OF KINEMATIC DATA

The pioneers in quantitative locomotion analysis, Braune and Fischer and Bernstein, used the graphic, i.e., the finite difference, method for derivative assessment of kinematic data. Viewed today, these methods are inadequate. However, with respect to the achieved spatial resolution, the measurements of these researchers were performed with high quality and precision. Woltring stressed the need for repeated analyses and processing of their measurement results by newer methods.

   Graeme A. Wood described and compared numerical procedures of smoothing and differentiation of human locomotion kinematic data which had been applied until his time.[131] His classification is used for graphical method, finite difference techniques, approximations by the least squares polynomial, spline functions, digital filtering, and Fourier analysis. The survey also included references.[134–139]

Graphical Methods: These are subjective methods in which the operator draws a smooth curve (line of best fit) approximately through the graphically presented experimental displacement data in the function of time and then measures the slope of a tangent on that curve in certain points. Derivative values so obtained represent velocities. Then, using the same procedure, a curve may be drawn through them and by measuring the slope of tangent on it, accelerations in certain time points can be determined. Such subjective and error-prone methods have been surpassed by the use of the computer. (The California Group of researchers, however, used graphoanalytical methods in their work.)

Finite Difference Techniques: If we denote some kinematic displacement function with $y = f(t)$, then the derivative $f'(t)$ of this function in some point (t_i) is defined as:

$$dy/dt = f'(t_i) = \lim_{h \to 0} \frac{f(t_i + h) - f(t_i)}{h} \qquad (4.22)$$

which represents a slope of the corresponding line segment. With $h \to 0$, the slope of this segment approximates the slope of a tangent in this point (t_i) better. For finite time intervals, a better approximation of derivative in t_i is the slope of another line segment:

$$dy/dt \cong f'(t_i) = \frac{f(t_i + h) - f(t_i - h)}{2h} \qquad (4.23)$$

Here $f(t_i + h) - f(t_i - h)$ represents the so-called first central difference.

   The second derivative in point t_i, acceleration, is obtained similarly by using central differences:

$$d^2y/dt^2 \cong f''(t_i) = \frac{f(t_i + h) - 2f(t_i) + f(t_i - h)}{h^2} \tag{4.24}$$

Wood[131] analyzes the use of such equations showing that they represent approximate values of the expressions obtained by the development of the analytic function f(t) in Taylor's series and concludes that, in general, when data contain a lot of noise, finite difference equations are not suitable for application. A better procedure is to first smooth kinematic data. It is even better to smooth them graphically and not directly apply numerical methods, like Bessel's[3] or Newton's formula. Pezzack et al.[134] suggested the use of the first central difference formula only after the application of a smoothing digital filter. Although traditionally data should be smoothed first, it is suggested that *a posteriori* smoothing of derivative curves determined by the finite difference method might yield acceptable results.

Approximations with Least Squares Polynomials: If one denotes experimental kinematic data with:

$$f(t) = p(t) + \xi \tag{4.25}$$

(ξ denoting error in empirical data), then the idea of applying the least squares method is that the data are fitted with a polynomial of not m-th, but of a lower order (m = n − 1, where n denotes number of data points), where it is assumed that biological function is approximated and noise is ignored. The expression for this lower order polynomial might be developed as an analytical curve of best fit. This is the standard method of least squares which requires minimizing the sum of squares of deviations S.

S is defined by the expression:

$$S = \sum_{i=1}^{n} [y_i - p(t_i)]^2 \tag{4.26}$$

For the m-th grade polynomial, this expression may be represented as:

$$S = \sum_{i=1}^{n} [y_i - a_0 - a_1 t_i - a2\, t_i^2 - \ldots - am\, t_i^{m-1}]^2 \tag{4.27}$$

For this function to be a minimum, partial derivatives $\partial S/a_0$, $\partial S/a_1$, . . . $\partial S/a_m$ have to be equal to zero. So, m equations (normal equations) are obtained which, when solved simultaneously, provide coefficients a_0, a_1, . . . , a_m for fitting the m-degree polynomial by applying the method of least squares to the set of empirical data. Once the expression for the polynomial is determined, time derivatives may easily be obtained by analytical differentiation of the approximate function.

When this procedure is applied to a small number of data (3 to 9), then it is a local approximation technique called moving average technique.

It is also a possibility that, in spite of the form: $a_0 + a_1 t + a_2 t^2 + \ldots + a_m t^m$, the function takes the form:

$$p(t) = a_0 p_0(t) + a_1 p_1(t) + a_2 p_2(t) + \ldots + a_m p_m(t) \qquad (4.28)$$

The practical advantage in the application of this method is that when m is incremented, only the new coefficient (and not m + 1 of them) must be calculated and the solution for the higher order polynomial is generally more stable. Often, Chebyshev polynomials are used: $p_m(t) = \cos (m \cos^{-1})$.

The shortcoming of methods which fit the polynomial is the inability of these functions to provide adequate approximations in time intervals of different complexity in one series of data. There is a need, therefore, for functions that may approximate data in different time intervals with different curving (which led to the use of spline functions).

Spline Functions: The spline function consists of more polynomials, all of which are low grade m, which are pieced together in points called knots (x_j, j = 1,2, . . . n) and connected so as to give a continuous function g(t) with m − 1 continuous derivatives. When m = 3, which is common in the practical application of this function, the resulting cubic spline function consists of n − 1 cubic polynomials, each taking the form:

$$g(t) = p_j(t) = a_j + b_j(t) + c_j(t)^2 + d_j(t)^3 \qquad (4.29)$$

which cover the interval $x_{j-1} \le t < x_j$ and satisfy the continuity condition

$$p_j^k(x_j) = p_{j+1}^k(x_j); \ (k = 0,1,2; j = 1,2, \ldots ,n) \qquad (4.30)$$

where p_j^k denotes the k-th derivative of the j-th segment of a polynomial. The condition where the function has m − 1 continuous derivatives secures its smoothness, but as opposed to the global polynomial, its piecewise nature enables it to adapt quickly to changes in curvature.

In the 1960s, the spline function proved to be the smoothest among all the functions for fitting N data into specified margins (Greville 1969, according to Reference 131). In practice, it is necessary to specify the degree of spline, the required accuracy of approximation (least square criterion), and the number and location of knots. The last one may, in practice, represent a problem. When there are enough data points (10 to 15 between each extreme and/or inflection point) and accuracy is relatively well known, the Reinisch method may be used. In this method, knots are put on each t_i, connecting N − 1 polynomial together in such a way which secures a smoothness integral for cubic spline defined as:

$$Q = \int_{t_1}^{t_N} [g''(t)]^2 \, dt \qquad (4.31)$$

is minimized subject to a least squares constraint condition

$$\sum_{i=1}^{N} \left[\frac{g(t_i) - y_i}{\delta y_i} \right]^2 \le S$$

where δy_i denotes standard measurement errors and S is a parameter which determines the degree of smoothing. The advantage of this method is that it is not necessary to decide where knots should be. This is also a method of choice for differentiation and integration, since it treats all points in the same manner.

Wood concludes that spline functions may be effectively applied and that they give good approximations to biomechanical data. Their extreme flexibility and pronounced local properties make them suitable to biomechanical applications but during application, the proper degree of function and the choice of conditions of least squares has to be taken into consideration.

Digital Filtering: Pezzak et al.[134] compared the results of smoothing kinematic data by using a digital filter, after which an estimation of the second derivative followed by using the finite differences method, with values of acceleration obtained directly. The application of a second order Butherworth filter proved to be the best choice after which came calculation of first order finite differences, while direct applications of finite difference techniques or approximations with Chebyshev polynomial resulted in nonacceptable results.

The formulation of nonrecursive filters often contains a great number of terms, so they are expensive computationally. Gustaffson and Lanshammar,[107] therefore, developed an iterative filtering procedure which used recursive and nonrecursive differenting filters. The nonrecursive filter was made to minimize the expected value of mean quadratic error in the derivative to be estimated, under the condition that k-th order estimates are unbiased (i.e., there are no systematic errors due to the filter) when the input signal is the polynomial of a degree m for m < k. Typically, 51 filter members were required, so a recursive filter formulation was developed which reduced the number of arithmetic operations by a factor of 5. However, the recursive algorithm was potentially unstable, so an iterative procedure was applied to monitor error induced by instability, by comparing the chosen recursive filter with what was predicted by means of a nonrecursive filter. These authors suggested a method by means of which a corresponding required sampling period can be determined based on precision limits of the measured and differentiated data and the width of the signal frequency spectrum, which has implications on the realization of kinematic measurement.

$$\sigma_k^2 = \sigma^2 \, \Delta \, \frac{\omega^{2k+1}}{(2k+1) \, \pi} \tag{4.32}$$

where $k > 0$ denotes the degree of derivative ($k = 0$ relates to smoothing without differentiation), ω is the lower limit frequency of lowpass filter, σ is the S.D. of noise in "raw" samples, and Δ is sampling time. Lanshammar, under white noise conditions, suggests sampling higher than determined by the Nyquist criterion (Expression 3.7).

Fourier Analysis: Periodic function f(t) might be represented in the form of a Fourier series:

$$f(t) = a_0 + a_1 \sin (2\pi t/T) + a_2 \sin (4\pi t/T) + a_3 \sin (6\pi t/T)$$
$$+ \ldots + b_1 \cos (2\pi t/T) + b_2 \cos (4\pi t/T) + b_3 \cos (6\pi t/T) \quad (4.33)$$
$$+ \ldots = a_0 + \sum_{j=1}^{n} [a_j \sin (j2\pi t/T) + b_j \cos (j2\pi t/T)]$$

This analysis may be applied prior to the application of a digital filter. The spectrum may also be estimated via FFT. By applying spectral analysis it has been determined that, in practice, kinematic signals of cyclic movements contain significant components up to the 7th harmonic. In children, however, gait is kinematically more complex. Apart from having to be applicable in the analysis of movement structures, such a representation may also be used as a direct means of obtaining time derivative information.

Wood cites equations, according to Cappozzo:

$$f'(t) = \sum_{j=1}^{n} j2\pi/T \, [-a_j \sin (j2\pi/T) + b_j \cos (j2\pi/T)]$$

$$f''(t) = \sum_{j=1}^{n} (j2\pi/T)^2 \, [a_j \cos (j2\pi/T) + b_j \sin (j2\pi/T)] \quad (4.34)$$

and also mentions the Hatze approach, by means of the optimally regularized Fourier series.

Comparison of Procedures: The most often used methods of kinematic signal processing in the field of human motion analysis, according to Wood,[131] were splines and digital filtering, and this holds true today. Several methods give acceptable results, while methods of finite differences, the simple moving averages method, and global polynomial approximation were considered to be inadequate by Wood. Digital filtering techniques are naturally adapted to the problem of noise separation, but their limit is that temporal data must be equidistant, and so they do not give an analytical function. Fourier analysis and splines satisfy these conditions, but Fourier analysis also requires data to be equidistant.

While Fourier analysis is ideally suited to the analysis of periodic data and gives the most efficacious way of storing of details of time series by means of a small number of coefficients, spline functions are very useful for biomechanical analyses. D'Amico and Ferrigno[138] compared the algorithm called the generalized cross-validatory splines algorithm (GCVC)[130] (Murphy and Mann 1987, according to Reference 138), the best-known and most-efficient automatic algorithm for smoothing of biomechanical data, with LAMBDA, a model-based derivative assessment algorithm. Comparisons have shown that the overall accuracy of LAMBDA is superior or similar to GCVC, while the latter requires a significantly longer time for execution. Furthermore, LAMBDA is insensitive to the number of samples, making it also possible for short data samples to be calculated (shorter than 40 samples) and minimizing distortions at the edges.

In treating the problem of smoothing and differentiation techniques of 3D data, Woltring[136] agreed with Wood and D'Amico. With regard to characteristics of the locomotor process, it is not clear if it is better to apply the differentiation procedure first to 1D data and then proceed to calculation of 3D, or vice versa. In gait, for example, the sinusoidal movements of various frequencies are determined, which can be seen in the results of gait studies by the Berkeley Group and Stokes et al., especially when discussing determinants of gait with regard to pelvic motion.

Mottet et al.[139] suggested a nonlinear method of smoothing the median called 7RY, suitable for application to kinematic signals of displacement prior to their differentiation. This method is especially suitable for the elimination of aberrant points in a signal.

As a final illustration for the application of measured kinematic signals of human movement in the pursuit of the inverse dynamic approach, a graphic representation of numerically estimated moments of forces in three lower extremity joints during walking is given, taken from the work by David Winter, an exceptionally clinically oriented researcher (Figure 4.35). Measurement data were obtained using a three-camera system: two located sagitally and one located frontally. The mechanical body model had nine segments: feet, lower leg, upper leg, pelvis, trunk, and head. Calculation and representation of this kind fulfill the goal of the inverse dynamic approach, and results, such as those shown, are a standard part of kinematic measurement systems applicable in research and clinical work.

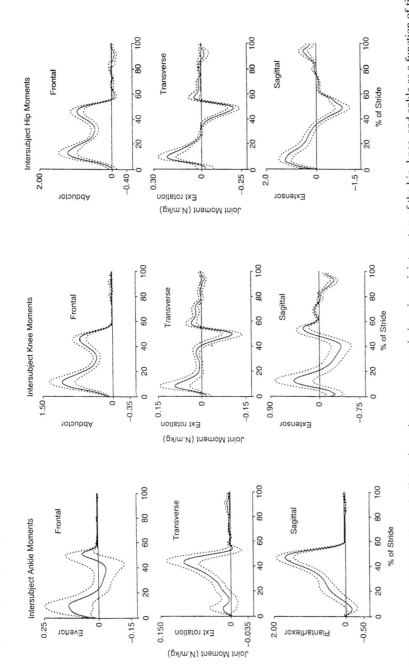

FIGURE 4.35 Numerically (mathematically) estimated moment curves in imaginary joint centers of the hip, knee, and ankle as a function of time during walking in a healthy individual: the result of an inverse dynamic approach. The measurement encompassed nine subjects, each performed five trials, while the kinetic values were calculated as "ensemble averages" of the group. (From Winter, D.A., Eng, J.J., and Ishac, M.G. 1996. *Human Motion Analysis*, G.F. Harris and P.A. Smith, Eds., New York: IEEE Press, 71–83. With permission.)

5 Measurement of Kinetic Variables

Kinetic (dynamic) movement quantities encompass forces and moments of force that are developed during movement: these are forces and moments between a body and its surroundings and internal forces and moments (Section 2.1). This chapter presents the procedures of measurement, processing, and interpretation of the total (resultant) forces and moments that develop due to contact between the body and the ground during support phases of locomotion ($\mathbf{F}_m$ and $\mathbf{C}_m$ in Figure 2.1) (Section 5.1). Historical development and a description of recent designs of measurement methods of pressure distribution between the body and the ground are also covered (Section 5.2). In both methods, data presentation and visualization of clinical and diagnostic value are presented.

If a measurement procedure is to be noninvasive, the measurement of the remaining (internal) kinetic quantities of human movement is practically impossible. This (unlike in animal experiments) is feasible only in special cases. Possibilities are telemetric measurement of distribution of mechanical loading in a hip joint using a special measurement device implanted in the artificial hip or *in vivo* measurement of force in the Achilles tendon.[141] Data of this kind, when available, are exceptionally useful, since they make it possible to validate the results of mathematical estimations of forces and moments obtained through the inverse dynamic approach.

5.1 GROUND REACTION FORCE MEASURING PLATFORMS

These devices enable measurement of the total force vector ($\mathbf{F}_m$ in Figure 2.1) manifested in various locomotor activities during contact between the subject's body (typically the foot) and the surface (floor) in which the device is embedded. Also, the device usually gives the moment of force vector ($\mathbf{C}_m$) as well as planar coordinate values x and y of the point of center of pressure R_n as its output. These measurement quantities may be displayed as time curves.[142] As such, the device is generally applicable in locomotion study, healthy or pathological. Besides being used in dynamic phenomena such as gait and running, force platforms (force plates) may also be used in measurements of approximate static body postures since, because body support via the feet is nearly fixed, measurement signals are a consequence of movement of the body's center of mass. This is exploited, for instance, when testing the vestibular apparatus from a neurological and otorhinolaryngological standpoint, in general, when examining postural stability and balance.

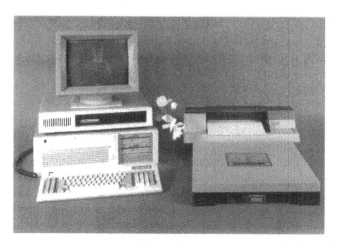

FIGURE 5.1   Ground reaction force measuring platform.

The instrument's contact area is a rectangular plate usually 60×40 cm in size (various other special designs of larger platforms are also possible) embedded in a firm, substantial base (Figure 5.1). The platform's surface must be at ground level, possibly covered by carpeting to enable truly noninvasive measurements (so the subject is not aware of having stepped onto the platform). The measurement room must be well equipped. There must be a track about 10 m in length for gait measurements and an even larger corresponding space for measurement of running, take-offs, etc. Moreover, it is best to position a chain of several platforms one after the other in the laboratory, thereby enabling measurement of successive strides in gait or running. However, due to cost considerations, this situation is rare.

During construction, measurement force transducers are positioned on and incorporated within the device. Depending on the kind of transducers and construction of the device, transducers must be positioned to attain selective sensitivity of the instrument when forces and moments of force are applied in all three spatial directions. An appropriate frequency response of the system is required, with resonant frequency of the subject-platform system reaching more than 200 Hz and with sufficiently low cross-talk between channels. The construction must ensure that force, i.e., moment values measured be independent of the place of application at the plate's surface.

The two most widely applied measurement instruments, the strain gage-based platform (Section 5.1.1) and the platform with piezoelectric transducers (Section 5.1.2), will be described.

5.1.1   THE STRAIN GAGE TRANSDUCER-BASED PLATFORM

The name given to this device is from the force transducers used, strain gages. The construction of a strain gage implies electrically conductive material, specially designed, that is affixed to a certain material body under the influence of force, so that it strains or stresses in concordance with the strain or stress of this body. The strain gage dates back to the late 1930s and was developed independently at MIT and Cal Tech. The layout of the sensor foil described here originated in the 1950s and was a by-product of printed circuit technology.

5.1.1.1 The Strain Gage Sensor

The principle of strain gage transduction is based on the phenomenon of mechanical strain. Mathematical development according to Božičević.[143] When a force acts on a body, it tries to accelerate it in the direction of its action, i.e., force causes movement of a body. When forces are balanced, they load the body, causing deformation—stress or strain. Force acting on the unit area of a rigid body is called strain. Begin with the assumption that the electrical resistance of the conductor, of which the gage is made, is directly proportional to its length and inversely proportional to its transverse diameter:

$$R = \rho \cdot l/A \; [\Omega] \tag{5.1}$$

where ρ denotes specific resistance, l denotes length, and A denotes area of wire cross-section.

If, therefore, the conductor, i.e., wire, is subjected to mechanical loading—stress or strain—some of its dimensions will change, and so its electrical resistance will also change. The material must have enough compliance, and its mechanical deformation is accompanied by a corresponding change in electrical resistance. Accordingly, a strain gage measures the relative deformation of the body in the vicinity of a certain reference point.

Denoting deformation with $\varepsilon = \Delta l/l$, the values in Equation 5.1 become:

$$l(\varepsilon) = l_0 (1 + \varepsilon)$$

$$A(\varepsilon) = A_0 (1 - \mu\varepsilon)^2 \tag{5.2}$$

$$\rho(\varepsilon) = \rho_0 (1 + \beta_p \varepsilon)$$

Here, l_0 and ρ_0 are values in an unloaded state, μ is the Poisson coefficient, and β_p is the coefficient of change of resistance due to strain.

The resistance of the loaded wire is

$$R = \rho(\varepsilon)l(\varepsilon)/A(\varepsilon) \tag{5.3}$$

By means of the expression above, the increase of the resistance value with change in deformation $d\varepsilon$ may be determined:

$$\frac{dR}{d\varepsilon} = \frac{\rho}{A}\frac{\Delta l}{\Delta\varepsilon} - \frac{\rho l}{A^2}\frac{\Delta A}{\Delta\varepsilon} + \frac{1}{A}\frac{\Delta\rho}{\Delta\varepsilon} \tag{5.4}$$

By substituting expressions for $\Delta l/\Delta\varepsilon$, $\Delta A/\Delta\varepsilon$, and $\Delta\rho/\Delta\varepsilon$, the following is obtained:

$$\frac{1}{R_0}\frac{dR}{\Delta\varepsilon} = 1 + 2\,\mu(1 - \mu\varepsilon) + \beta_p \tag{5.5}$$

$$d\varepsilon = dl/l_0$$

Therefore, the sensitivity of the strain gage is

$$S = \frac{dR/R_0}{dl/l_0} \tag{5.6}$$

and since the product $\mu\varepsilon$ may be neglected ($<< 1$):

$$S = 1 + 2\mu + \beta_\rho \tag{5.7}$$

Sensitivity S is often denoted as k and called the k-factor (or gage factor), which is a function of the material, and its value amounts to around 2 for metals.

Strain gages are made in the form of wires or foil— foil strain gages which are manufactured by engraving. They are usually 0.02 to 0.05 mm thick and are designed in complex geometrical shapes, such as those depicted in Figure 5.2. In this way, their working span is increased and they are adapted to specific applications. The geometric shape of a gage defines its direction of sensitivity. According to physical principles,

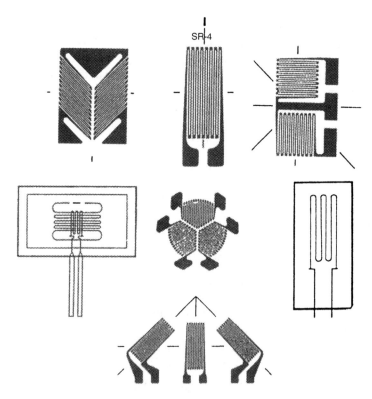

FIGURE 5.2 Various strain gage designs in the form of geometrically formed foil made of conductive material. Foil must be positioned on the supporting surface, to achieve sensitivity to the application of force in the desired direction.

it is a passive transducer since the action of an external force causes a change only to the transducer's features (of mechanical form and electrical resistance) and only with the use of an external voltage source (DC voltage source) is an indication of change realized. (A passive-active transducer distinction could also be made with regard to another criterion, position in the measuring bridge, according to which active sensors are those that are exposed to deformation.[143])

Design and layout of strain gages are the result of a compromise between the requirement for flexibility, with the goal of attaining the highest possible degree of sensitivity, and the requirement for rigidity, with the goal of realizing the highest possible characteristic frequency.

Since the resulting resistance changes in a strain gage are very small, gages are connected into a Wheatstone bridge circuit, where they represent elements: electrical resistors R_{1-4} (Figure 5.3). The mathematical development applied is according to Barnes and Berme.[144,145] E denotes the output voltage. In equilibrium, with $E = 0$, $R_1/R_2 = R_4/R_3$. Further:

$$i_{BAD} = V/(R_1 + R_4)$$

$$i_{BCD} = V/(R_2 + R_3)$$
(5.8)

Next:

$$E_{AC} = E_{AB} + E_{BC} = E_{AB} - E_{CB} = i_{BAD} R_1 - i_{BCD} R_2$$
(5.9)

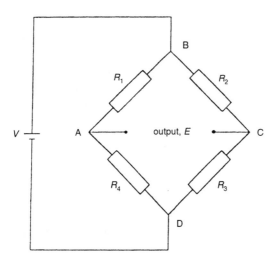

FIGURE 5.3 Strain gage sensors, of corresponding resistance values $R_1 - R_4$, connected in a Wheatstone bridge circuit. (From Berme, N. 1990. *Biomechanics of Human Movement: Applications in Rehabilitation, Sports and Ergonomics,* N. Berme and A. Cappozzo, Eds. Worthington, OH: Bertec Corporation, 140–148. With permission of N. Berme/Bertec Corporation.)

By substituting Equation 5.8 in Equation 5.9:

$$E_{AC} = [R_1/(R_1 + R_4) - R_2/(R_2 + R_3)] \, V \qquad (5.10)$$

If leg resistances $R_1 - R_4$ in an unloaded state are equal to R, and the bridge is wired such that R_1 and R_3 increase by ΔR, while R_2 and R_4 decrease by ΔR, then:

$$\Delta E = \frac{\Delta R \, R}{(R + \Delta R)^2} \, V \qquad (5.11)$$

where ΔE denotes the change in output value. The output voltage is linearly proportional to the excitation voltage, but nonlinear with the change in gage resistance. However, as $\Delta R << R$, the last expression can be linearized, to obtain:

$$\Delta E = (\Delta R/R) \, V \qquad (5.12)$$

Gages are designed in such a way that, when maximally loaded, a 0.1% resistance change typically occurs. With $\Delta R/R$ equal to 0.001, Equations 5.11 and 5.12 show that, at full scale deflection, the transducer output will have a 0.2% deviation from linearity.

In most transducer designs, each leg is made so that initial gage resistances are equal and all four gages contribute to force measurement. In this way, the sensitivity of the transducer is increased and simultaneously a compensation of thermally induced strain is achieved. Strains from the opposing legs are added, while adjacent ones are subtracted, thus neutralizing the temperature effect. With suitable gage placement, it is possible to isolate and measure any one of the combined loads separately. Conversely, any one general mechanical load (force and/or moment) may be decomposed into its Cartesian components. Figure 5.4 shows a cylindrical beam subjected to a combined mechanical loading state that includes a bending moment around the x-axis, a y-shear force component with tangential action, and a vertical or axial z-force component. The moment shown may represent either a vertical force acting at some distance along the y-axis or a pure couple (free moment) applied as shown. Figure 5.5 shows an arrangement, i.e., sensor element layout, where the bridge output is only sensitive to a vertical, axially acting force. There are two options when measuring shear forces. Figure 5.6 shows shear force measurement by comparing the bending moment deformation of the column at two different levels along its long axis. Figure 5.7 shows an arrangement utilizing gages placed at 45° angles with a long axis of the beam which measures the shear force directly from the shear deformation of the beam. (With appropriate alignment of the gages, the effects of any torsion that causes shear deformation can be canceled, so the output is the function of only shear force.)

All four bridge elements need not be on the same transducer element. Also, more than one gage resistance can be placed in the same leg of the Wheatstone bridge. Furthermore, the output of more than one bridge can be coupled to give the sum effect of a force component acting on two or more transducer elements (Figure 5.8). When

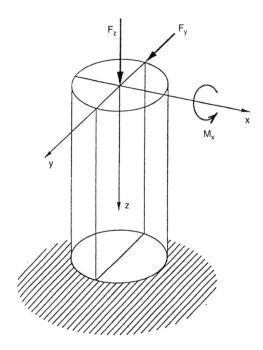

FIGURE 5.4   A cylindrical beam fixed at one end and subjected to combined loading at the other. (From Berme, N. 1990. *Biomechanics of Human Movement: Applications in Rehabilitation, Sports and Ergonomics,* N. Berme and A. Cappozzo, Eds. Worthington, OH: Bertec Corporation, 140–148. With permission of N. Berme/Bertec Corporation.)

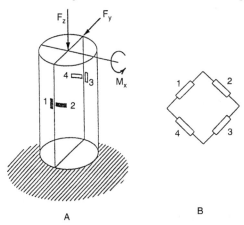

FIGURE 5.5   Axial force measurement. (A) Gage locations on the beam; (B) gage connection in the Wheatstone bridge. (From Berme, N. 1990. *Biomechanics of Human Movement: Applications in Rehabilitation, Sports and Ergonomics,* N. Berme and A. Cappozzo, Eds. Worthington, OH: Bertec Corporation, 140–148. With permission of N. Berme/Bertec Corporation.)

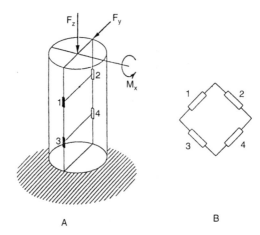

FIGURE 5.6   Shear force measurement using the bending moment difference method. (A) Gage locations on the beam; (B) gage connections in the Wheatstone bridge. (From Berme, N. 1990. *Biomechanics of Human Movement: Applications in Rehabilitation, Sports and Ergonomics,* N. Berme and A. Cappozzo, Eds. Worthington, OH: Bertec Corporation, 140–148. With permission of N. Berme/Bertec Corporation.)

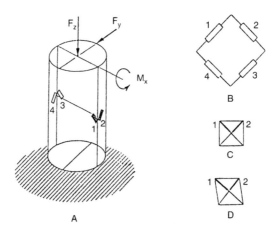

FIGURE 5.7   Shear force measurement using 45° gage orientation. (A) Gage locations on the beam; (B) gage connections in the Wheatstone bridge. Gages 1 and 2 in the (C) unloaded and (D) loaded state. (From Berme, N. 1990. *Biomechanics of Human Movement: Applications in Rehabilitation, Sports and Ergonomics,* N. Berme and A. Cappozzo, Eds. Worthington, OH: Bertec Corporation, 140–148. With permission of N. Berme/Bertec Corporation.)

bridges are split or combined, the gage sensitivities must be matched in order to maintain the independence of transducer output on load position. These rules apply to all three of the loading cases described above. This is the basis for strain gage-based force plate transducer design. The bridge output signal is fed into the amplifier and after amplification it is available for further processing, A/D conversion, etc.

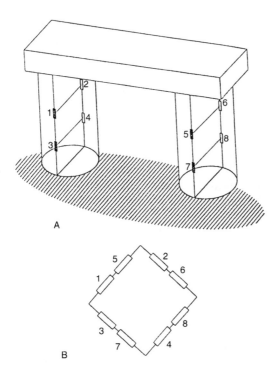

FIGURE 5.8 Combination of two shear bridges in one. (A) Gage locations on individual beams; (B) gages connected into one bridge. (From Berme, N. 1990. *Biomechanics of Human Movement: Applications in Rehabilitation, Sports and Ergonomics,* N. Berme and A. Cappozzo, Eds. Worthington, OH: Bertec Corporation, 140–148. With permission of N. Berme/Bertec Corporation.)

5.1.1.2 Platform Construction

A strain gage transducer-based platform usually has four rectangularly arranged load transducers separating two plates. A certain loading pattern will induce reactions in all four transducers. In order to obtain output for each respective channel of the total device output (forces will be denoted with F_x, F_y, F_z, and moments will be denoted with M_x, M_y, M_z), outputs of each transducer are added together and principles of summation and superposition are used as previously described (see Figure 5.8). In this way, each output channel represents a direct measure of corresponding force, i.e., moment.

AMTI (Advanced Mechanical Technology, Inc.) in Newton, Massachusetts, has been marketing its strain gage-based force plate since 1976. This was the first commercial platform constructed to use strain gages and was intended for gait analysis in the biomechanics laboratory at the Children's Hospital in Boston. Since then, with the appearance and development of microcomputers, corresponding PC and user software support have been offered for a comprehensive solution of the measurement system. This instrument is used in many scientific research laboratories and clinical environments.

In a quasi-static case, horizontal forces and torques may be neglected. It is presumed that the platform is loaded only in a vertical direction. To satisfy the equilibrium condition, the following must hold (*AMTI Newsletter*):

$$\Sigma F_z = 0 = R - F_1 - F_2 - F_3 - F_4$$

$$\text{or } R = F_1 + F_2 + F_3 + F_4 = F_z$$

(5.13)

where F_z denotes the platform output signal, equal to the electrical sum of the four internal reaction forces. This signal is a measure of the resultant vertical force applied. The next requirement for static equilibrium is that the sum of moments around any one axis be equal to zero:

$$\Sigma M_x = 0 = R\,b + (F_1 + F_4 - F_2 - F_3)\,(L/2)$$

or

(5.14)

$$R\,b = (F_2 + F_3 - F_1 - F_4)\,(L/2) = M_x$$

M_x is the corresponding device output signal which equals the electrical combination of four reaction signals multiplied by the factor representing the effective length of the lever, one half of the distance between the sensors, whose value is constant, determined by the calibration procedure. Symbols a and b denote coordinates of the point of center of pressure in x and y directions, respectively. By substituting R from Expression 5.13 and dividing it with this value:

$$b = \frac{(F_2 + F_3 - F_1 - F_4)\,L}{(F_1 + F_2 + F_3 + F_4)2} = \frac{M_x}{F_z}$$

(5.15)

By analogy:

$$a = \frac{(F_3 + F_4 - F_1 - F_2)\,L}{(F_1 + F_2 + F_3 + F_4)2} = \frac{-M_y}{F_z}$$

(5.16)

In this way, the coordinates of the center of pressure on the platform (point R_n in Figure 2.1) are calculated in the quasi-static case.

In the majority of biomechanical applications, the values of horizontal components of moments of force M_z and M_y are proportional to the location of the center of pressure (the expressions stated) and the vertical moment includes a moment of torsion around the vertical axis. In general, each moment measured by the platform includes the moment of force component (the product of force and vertical distance from the point to the force vector) and the free moment (couple) (two noncolinear parallel forces acting in opposite directions). In general, the platform scheme is given in Figure 5.9. The following is true:

$$M_z = -F_x\,b + F_y\,a + T_z$$

(5.17)

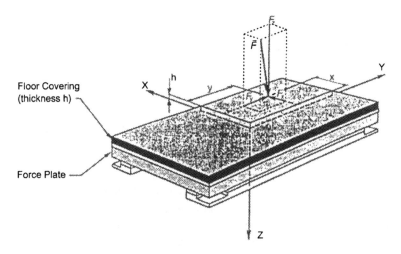

FIGURE 5.9   Ground reaction force measuring platform. Floor covering is at a distance h from the platform's origin. (From Barnes, S.Z. and Berme, N., 1995. *Gait Analysis: Theory and Application,* R.L. Craik and C.A. Oatis, Eds. St. Louis, MO: C.V. Mosby, 239–251. With permission.)

Products F_x b and F_y a denote moments of force and T_z is the free moment. (The negative sign is due to the right coordinate system being used.) The free moment T_z is the quantity sought in biomechanical studies and is calculated from the above expression.

By analogy, for horizontal moments:

$$M_z = F_z\, b + F_y\, h + T_x$$
$$M_y = -F_z\, a - F_x\, h + T_y$$

(5.18)

In the majority of biomechanical applications, the free moments T_x and T_y are equal to zero. With the subject standing or walking on the platform, the only way for him to generate a free moment around the horizontal axis would be to stick to the platform with his foot, so as to twist the upper surface around the x or y axis. Otherwise, there are no free moments T_x and T_y so the expressions mentioned may be employed to calculate the coordinates of center of pressure:

$$a = -\frac{M_y + F_x\, h}{F_z}$$

(5.19)

$$\Omega b = \frac{M_x + F_y\, h}{F_z}$$

Expressions 5.15 and 5.16 are special cases of Expression 5.19.

The platform's dynamic features, meaning the manner in which it reacts to time-varying forces and moments which act upon it, depend on the transducers' features, the construction of the device, and how it is fixed to the ground surface. The system's lowest resonant frequency is critical. Limiting considerations to a vertical loading direction only, the lowest resonant frequency will depend on:

- Vibrations due to the bending of the upper plate
- Vertical vibration of the upper plate treated as a rigid body, where force transducers act as springs
- Vibration of the platform as a whole possibly due to poor fixation to the ground

In practice, if the base is not affixed to the ground, the lowest frequency will be 220 Hz (AMTI 1984). When the base is properly fixed, the limiting frequency is due to the bending of the upper plate, which amounts to 398 Hz in this particular device. Since the human body is not a rigid mass, its collision with the platform does not lower this value significantly. In platforms equipped with rigid sensors and properly fixed to the ground, the limiting resonant frequency is a function of the bending of the upper plate (about 500 Hz in the AMTI OR6–5). AMTI reports the following technical characteristics of the device (series OR6-6-2000, dimensions of the contact surface 50.8 × 46.4 cm): sensitivity, 0.08 μV/VNm for F_z (0.34 for F_x and F_y) and 1.68 μV/VNm for M_z (0.80 for M_x and M_y). The working range is 8900 N for F_z, 4450 N for F_x and F_y, i.e., 1130 Nm for M_z and 2260 Nm for M_x and M_y. Nonlinearities are ± 0.4% of the full scale deflection for all three force components. Hysteresis is 0.4% of the full scale deflection for all three force components. Several instrument designs are offered (a transparent one among others, which may be of interest for some applications).

In the Biomechanics Laboratory at the Faculty of Physical Education in Zagreb, a custom-designed strain gage-based force plate is used. The dimensions of its contact surface are 40 × 40 cm and its mass is 20.1 kg. It is fixed into a metal-wooden frame. Each one of the three orthogonal total force components acting on the platform, F_x, F_y, and F_z, is measured by means of a separate Wheatstone bridge. The bridges give output signals which are fed, through coupler modules, to preamplifiers of the Beckman Dynograph device. Each coupler module provides excitation voltage (generated in the device) at its output. The module further enables precise balancing of the bridge by connecting proper resistors (the range of strain gage resistance values allowed is 90 to 5000 Ω) and also by connecting the resistors in parallel to one branch of the bridge, the simulation of force value with the goal of providing calibration. Ink pen paper printout of signals is enabled, as well as A/D conversion and storage in a PC (Section 3.2). Major technical characteristics of the device are: maximum sensitivity of 2.45 N/mm for F_z, 1.44 N/mm for F_x, and 1.36 N/mm for F_y (50 mm equals to 2.83 V at the output). The working range is ±4 kN for all three components. A number of measurements have been realized by using this instrument; the results of some are given in this and subsequent chapters.

There is one common problem encountered in strain gage-based and other platforms: cross-talk between channels caused by nonidealities in device layout. Therefore, each particular instrument has to be appropriately calibrated and correction of the identified cross-talk has to be provided. This task can be achieved by using software solutions.

5.1.2 THE PIEZOELECTRIC TRANSDUCER-BASED PLATFORM

Besides strain gage-based platforms, another kind of instrument is also used for measuring locomotion. It is based on another physical principle of measuring forces and moments: the piezoelectric effect. It is a kind of active transducer, since the transduction of mechanical into electrical energy occurs without an external energy source. A device by the Swiss firm Kistler, based on this type of measurement sensors has widespread application and may be said to be a standard in biomechanical locomotion measurements. The first commercial platform manufactured by Kistler was in 1969. It was intended for gait analysis in the biomechanics lab at ETH in Zürich (J. Wartenweiler). Basic physical characteristics and specificities of piezo-electric measurement sensors will be presented.[145–149]

5.1.2.1 The Piezoelectric Effect

The piezoelectric effect (piezo, pressure) was discovered in 1880 by the brothers Pierre and Jacques Curie, while transducers based on this physical phenomenon were not in practical use until the 1940s. A feature of some materials of crystalline atomic structure, which when influenced by mechanical strain generate electrical potential, is of concern here; an electrical potential is created by the movement of charges along certain crystallographic axes. This effect is reversible, i.e., the application of electrical charge induces mechanical oscillations of the material. Electrical charge values are minute, of the order of magnitude of a pC, imposing high requirements on the layout of electronic amplifier circuits. Apart from this, stringent requirements are also placed on necessary isolation materials.

In spite of the fact that the principle of the electrometer amplifier had been known for a long time, electrometer tubes good enough for application were not constructed until the early 1950s. In the 1950s, W.P. Kistler patented the principle of the charge amplifier (description follows), but it did not gain practical importance until the early 1960s. A real breakthrough came with the invention of metal oxide semiconductor field effect transistors (MOSFETs), which replaced the electrometer tube. Also, the invention of isolation materials, such as Teflon and Capton, in the mid 1960s was of decisive importance. This meant the beginning of widespread application of piezoelectric measurements in many fields of engineering and other measurement applications.

Piezo crystals are characterized by high impedance, and, therefore, they are able to generate only very small currents. To induce a charge, only small mechanical deformation is needed (of an order of magnitude of μm); these transducers, contrary to strain gages, are very rigid. The intensity of the effect is expressed in pC/N. There are ten or so natural materials which may be used to manufacture transducers. Typical among them are quartz, characterized by excellent stability, Rochelle salt, and barium titanium. These materials are of high sensitivity and, as stated previously, of high rigidity, which is why their natural resonant frequency is about 30,000 Hz.

Quartz as a piezoelectric material: Quartz crystal is most often used and is the most suitable piezoelectric material available.[148] It is characterized by high isolation resistance, high mechanical strength, a high Young modulus (the modulus of elasticity in the longitudinal direction), and the absence of the pyroelectric effect and hysteresis, and it has extremely high linearity and excellent stability. Taking quartz as an example, various piezoelectric effects, such as the longitudinal, transversal, and shear effect, can be identified and used (schematically shown in Figure 5.10). Coordinate axes correspond to the crystallographic axes of quartz. The z axis is called the optical axis, and the x axis is called the electrical axis.

In the longitudinal and shear effects shown, the electrical charge is directly proportional to the total force applied and appears on the same surfaces of the quartz material which are mechanically loaded. Moreover, output charge is independent of size and form of the quartz element. The transversal effect differs significantly in that,

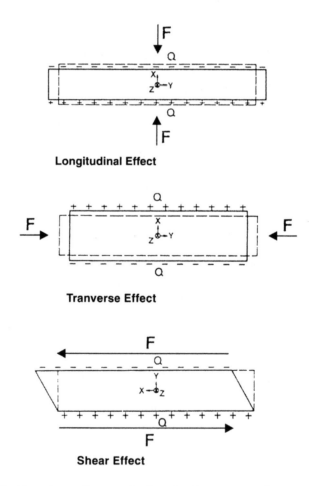

Longitudinal Effect

Tranverse Effect

Shear Effect

FIGURE 5.10   Crystallographic axes of a piezoelectric quartz crystal.

while the proportionality between the force applied and the electrical charge remains, the charge does not appear on the surfaces which are mechanically loaded, but on those surfaces normal to the x-axis, and is no longer independent of the form of the quartz element. The charge now depends on the mutual relationship between dimensions x and y, and so, by choosing a slimmer and longer quartz element, higher output may be obtained. Evidently, there is a limit since by making the element thinner, the permitted maximum bending load decreases. The transversal effect is used mostly for pressure transducers and force transducers of high sensitivity. Normal sensitivity for the longitudinal effect is 2.31 pC/N and is 4.62 pC/N for shear effect (each for one quartz element), while sensitivities above 500 pC/N can be achieved through the transversal effect.

Piezoelectric transducers can be designed for specific applications, i.e., modes of mechanical operation: compression, bending, or shear, which is, after all, a function of the geometry of the crystalline grid of the piezoelectric material. The property of quartz crystal, as a piezoelectric material, is convenient due to the possibilities of manufacturing multicomponent force transducers. The quartz material must be cut into discs. Depending on the direction of the cut, discs are made sensitive, either only to pressure (longitudinal effect) or only to shear, in one direction. When the disk is fixed between metal electrodes and is subjected to the action of force, the crystal material deflects and, proportionally, an electrical charge is generated. At the same time, voltage develops between the electrodes and is amplified. To measure one loading component, one pair of quartz discs is normally used. Using discs in pairs doubles the sensitivity and enables simple electrical contact through the middle electrode. By combining three pairs of discs, one pair for axial loading and two pairs for two orthogonal shear loadings, a three-component force transducer is obtained. These kinds of transducers have to be assembled with compressive mechanical preloading (prestress), holding the discs together, which prevents the crystal from being separated from the mechanical electrodes under the action of tensile loading.

Furthermore, it is also possible to design moment transducers by arranging a number of crystals cut in the y direction, sensitive to the shear force, and arranged in a circle with tangentially directed sensitivity axes to the shear loading. This kind of arrangement, fixed between two metal rings and with high mechanical preload, will measure rotational moment. In high-quality quartz cutting, cross-talk of $< 1\%$ is achieved.

The charge amplifier: As mentioned previously, crystal sensors have a high resistance output, so charge amplifiers are needed. The charge amplifier is a classical DC amplifier with capacitive feedback, with C_p being the capacitance of the feedback capacitor and C_{in} being the cable and amplifier capacitance acting in parallel to the amplifier's input.[143] The following expression holds:

$$u_{out} \cong -\frac{Z_p}{Z_{in}} u_{in} \qquad (5.20)$$

Z_{in}, an input impedance, is $-j/ \omega(C_A + C_u)$, where C_A is transducer capacity. Z_p is the value of feedback impedance, $-j/ \omega C_p$. So:

$$u_{out} \cong -\frac{C_A + C_u}{C_p} u_{in} \tag{5.21}$$

Amplifier input voltage may be expressed in the function of the quantity of charge generated by the piezoelectric transducer:

$$u_{in} = q/(C_A + C_u) \tag{5.22}$$

Transducer sensitivity is defined as $S = q/a$ (a denoting mechanical acceleration), and input voltage may be expressed as:

$$u_{in} = \frac{q}{C_A + C_u} \cdot \frac{a}{a} = \frac{S}{C_A + C_u} \cdot a \tag{5.23}$$

Furthermore:

$$u_{out} = -\frac{S}{C_p} \cdot a \tag{5.24}$$

With input voltage e_i and a negligibly small current i_g the following holds:

$$k\frac{dx_i}{dt} = -C_p \frac{du_i}{dt} \qquad u_i = \frac{k}{C_p} x \tag{5.25}$$

Therefore, e_i linearly depends linearly on displacement x. However, this is an ideal situation while, in practice, resistor R_p is added to the feedback capacitor and its value may be chosen between 10^{11} and 10^{14} Ω. It then follows:

$$u_i = k_0 \tau [1 - \exp(-t/\tau)]$$
$$k_0 = k/C_p \text{ [V/mm]} \qquad \tau = R_p C_p \text{ [s]} \tag{5.26}$$

The sensitivity of the charge amplifier used with the quartz crystal transducer ranges from 0.01 to 100 mV/pC. The frequency characteristic reaches from the DC to the value of an order of magnitude of 10^5 Hz. Nonlinearity is less than 0.1% and output noise is about 2 mV with short-circuited input. The characteristic frequency is around 1 kHz. Deflection during measurement is minimal. It exhibits extreme linearity and high stability with a broad working range (range/threshold around 10^6).

5.1.2.2 Platform Construction

Usually, four identical force transducers are used, each positioned in one corner of the platform (Figure 5.11). Moment values are deduced from the forces measured and from the relative positions of transducers in the platform, which is why there are eight output channels (and not six like in a strain gage-type platform). Twelve of the outputs from the four transducers are connected so that there are finally eight outputs.

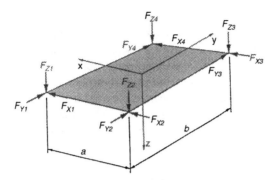

FIGURE 5.11 Twelve individual force components measured by a piezoelectric force plat-form. The eight platform outputs are $(F_{x1} + F_{x2})$, $(F_{x3} + F_{x4})$, $(F_{y2} + F_{y3})$, F_{z1}, F_{z2}, F_{z3}, and F_{z4}. (From Barnes, S.Z. and Berme, N. 1995. *Gait Analysis: Theory and Application*, R.L. Craik and C.A. Oatis, Eds. St. Louis, MO: C.V. Mosby, 239–251. With permission.)

There are four separate measurements of vertical force, two tangential components in the x direction and two tangential components in the y direction. In order to obtain six components of force and moment, data are further reduced to:

$$F_x = (F_{x1} + F_{x2}) + (F_{x3} + F_{x4})$$

$$F_y = (F_{y1} + F_{y4}) + (F_{y2} + F_{y3})$$

$$F_z = (F_{z1} + F_{z2}) + (F_{z3} + F_{z4})$$

$$M_x = [(-F_{z1} - F_{z2}) + (F_{z3} + F_{z4})]\, b/2$$

$$M_y = [(-F_{z1} + F_{z2}) + F_{z3} - F_{z4})]\, a/2$$

$$M_z = [(F_{x1} + F_{x2}) - (F_{x3} + F_{x4})]\, b/2 + [(F_{y1} + F_{y4}) - (F_{y2} + F_{y3})]\, a/2$$

(5.27)

When only compressive forces in the z-direction are applied to the platform, only the free moment in the x-y plane can be transferred to the platform (in analogy with considerations on the strain gage based platform). The point of application of the resultant force and the free moment can be calculated from the value of the force measured and the moment components expressed within a Cartesian coordinate system by using Expressions 5.17 and 5.19.

The Kistler platform provides measurement of the total vertical force and horizontal components of the force applied with an accuracy better than 1%, nonlinearity and hysteresis less than 1%, and sensitivity up to 0.05 Pa, in a working range typically 200 kPa for the vertical and ± 50 kPa for the horizontal force component. Rather good repeatability and long-term stable properties make this instrument standard application.

There are specific comparative differences between measuring platforms based on the piezoelectric effect and those using strain gage-type transducers. Since frequency response of piezoelectric systems to mechanical excitation is very high, these transducers are indispensable for certain special applications. However, the piezoelectric system, as an active system, does not allow only static measurements such as those by means of strain gage transducers, but quartz as a piezoelectric material in connection with a charge amplifier nevertheless offers the possibility of measuring approximately static phenomena that may last for a number of minutes or even hours. For biomechanical studies of human locomotion, this is completely satisfactory.

5.1.3 MOUNTING THE FORCE PLATFORM

The platform must be properly positioned on the floor and, if possible, covered by a carpet of some kind to enable as noninvasive a measurement procedure as possible. Naturally, a platform covering is not required when measurements of quasi-static body positions or of take-offs from a defined place on the ground are carried out. (There are situations in which the point of contact is spatially defined *per se*, e.g., in take-off for a long jump.) The device must be appropriately embedded, meaning possible vibrations of the device during loading and impact have to be minimized. This is achieved by building a rigid mounting surface of high mass in the floor and by fixing the platform to the surface.

AMTI provides recommendations for the proper embedding of the platform, the most important among them being the following. A flat mounting surface is required, with a tolerance of up to 0.05 mm and a corresponding rigidity of the frame. Because frame rigidity influences the platform's resonant frequency, it has to be as rigid as possible for technical data on the lowest resonant frequency to be fulfilled. If the laboratory facility is located on the upper floor of a building, possible vibrations of the complete supporting structure have to be considered. In some measurements, however, the system's full frequency response is not required, for instance, when measuring body sway, as will be seen. Further, AMTI offers some interesting possibilities of mobile, air-filled systems. In practice, the platform may be used even if it is not permanently embedded in a floor. In this case, at least an adequate walking path has to be designed, built of wood, to make the surface levels of the device and the surrounding floor equal.

5.1.4 APPLICATIONS OF THE FORCE PLATFORM

Certain typical measurement results, in the form of time curves and obtained by means of a force platform, in healthy and pathological locomotions are presented.

5.1.4.1 Walking and Running

Walking and running, the most natural human locomotion, have often been the subject of ground reaction force measurements. Saunders and colleagues[87] presented a comparative study of vertical ground reaction force signal components during stance phase in a healthy and unhealthy leg, specifically gait pathology resulting from a

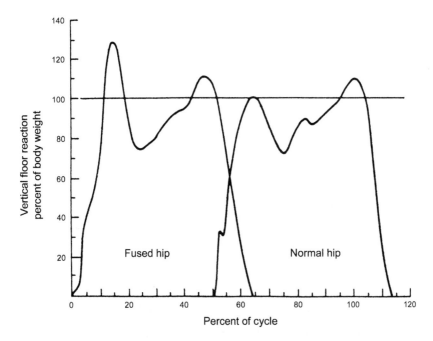

FIGURE 5.12 Differences in vertical ground reaction forces exerted by a healthy and injured leg, in an individual with a fused hip as a consequence of arthrosis. These abnormalities are reflected on the normal side, indicating an attempt in the normal hip to compensate for the loss of motion on the affected side. (From Saunders, M., Inman, V., and Eberhart, H.D. 1993. *J. Bone Jt. Surg.* 35-A(3):543–558. With permission.)

fused hip (Figure 5.12), as part of their comprehensive research into healthy and pathological gait. The two signal waveforms differ significantly (which will be commented on after presentation of the results of healthy gait).

Nilsson and colleagues[150] experimentally researched the biomechanical and neurophysiological features (by means of EMG) of functions of the locomotor system during gait and running, specifically the transition between these two modes of locomotion. Figure 5.13 illustrates corresponding ground reaction force signals from two healthy subjects, at four different walking and running speeds, measured on the Kistler platform. Subjects wore athletic footwear. While the first foot-ground contact in normal gait always occurs with the heel, in running this may not be the case. One of the subjects chosen characteristically made first contact with the heel (rearfoot striker, the subject PS). The other made first contact with the front part of the foot (forefoot striker, subject OS).

In walking (subject OS), the vertical ground reaction force component F_z shows a characteristic waveform with two maxima and a local minimum in between, occurring approximately halfway through the support phase. This instant marks the amortization of vertical body movement by knee joint activity (the third gait determinant, Section 4.2.1). Running is characterized by a different ground reaction force signal waveform. The local minimum is absent and the signal is "monophasic," with a more

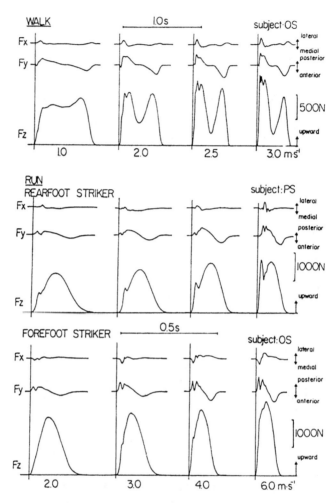

FIGURE 5.13   Mediolateral (F_x), anteroposterior (F_y), and vertical (F_z) ground reaction force component waveforms exerted by the right foot in walking (1.0, 2.0, 2.5, 3.0 m/s) and running (2.0, 3.0, 4.0, 6.0 m/s), with a rear- (PS) and forefoot (OS) strike, respectively. The body weight of subject OS was 790 N and PS was 720 N. Kistler measuring platform. (From Nilsson, J., Dissertation. On the Adaptation to Speed and Mode of Progression in Human Locomotion. Stockholm, Sweden: Karolinska Institute. With permission of *Acta Physiol. Scand.*)

or less expressed impact impulse (a function of the rigidity of impact: this impulse is more pronounced in rearfoot strikers). Nilsson presented general features of such measurement signals recorded in a healthy population. The vertical force peak is between 1 and 1.5 times the body weight values and in running is between 2 and 3 times the body weight value. The vertical signal peak due to impact, characteristic of rearfoot strikers, increases with speed, from 1.3 to 2.6 times body weight value. (In the above discussion, when "foot" is mentioned, it means "foot in footwear," not barefoot.)

To enable intersubject comparisons, as well as intertrial comparisons in the same subject, it is possible to express the time scale in such representations in percentages of the values of the duration of one cycle, while force amplitude can be represented in percentages of body weight.

Inspecting the horizontal force signal components shown, it may be observed that, in walking, the anteroposterior component (F_y) shows a small initial peak in the anterior direction, which changes into a posteriorly directed braking force, which changes into a propulsive one in mid-support. Rising phases of the first vertical and braking horizontal force sometimes have small superimposed peaks, primarily at high speeds. The mediolateral ground reaction force component (F_x) always shows a laterally directed peak at the instant of foot strike (i.e., the action force from the foot is directed medially), after which a medially directed reaction force generally follows, during a larger part of the support time.

In running, the anteroposterior force signal component shows the same braking and propulsive pattern as in gait. In both subjects, typical representatives of their groups, braking anteroposterior force manifests doubled peak value. Mediolateral force is a relatively complex waveform. Forefoot strikers always manifest an initial, medially directed reaction force peak, while rearfoot strikers mostly manifest a typical laterally directed peak value. In this respect, this second phenomenon is similar to walking, in which foot-strike always occurs with the heel first. There is a striking difference in the timing of anteroposterior and mediolateral force peaks between these two individuals. The first and second anteroposterior and mediolateral force peaks coincide at all speeds among forefoot strikers, but are clearly separated in time among the rearfoot strikers. It is clear that, under the two conditions, the impact of the foot will differ significantly, but Nilsson did not establish corresponding kinematic correlates in his research.

In gait, as well as in running, ground reaction force signals reflect an increase in the movement speed through an ever-larger increase in peak values and a shortening of signal duration. These measurement signals in healthy subjects are important for the interpretation of signals from pathologies. The example presented earlier of a fused hip signal, from measurements by the California Group, illustrates one form of pathological gait (see Figure 5.12). Deviations from a normal signal waveform may be observed on both sides. The remaining normal segments of the attacked extremity are only able to partially compensate for the loss of hip joint function. It should be remembered that pelvic rotation and pelvic tilt are significant factors for the decrease of vertical displacement of the body's center of gravity during walking. In the maximum elevation point of the body's center of gravity, the hips are in the middle position with regard to rotation. From this point, and until double support, when the feet are mutually the furthest apart during a stride, the pelvis rotates externally around the weight-bearing hip. This may be expressed as internal femur rotation with respect to the pelvis; it is of small amplitude, but an important movement. When the hip is diseased resulting in limitation of movement, it is kept in external rotation during the swing phase, to be in a position to enable small internal rotation which remains during the support phase and to enable pelvis rotation in a normal manner as far as possible.

Milinković[151] compared the features of ground reaction force signals in gait in healthy women and in those with unilateral lower leg amputations, with prostheses, by using the strain gage force platform. Seven female participants with unilateral lower leg amputations took part in the study, as well as ten healthy women who served as a control group. The groups were age-matched and the amputees were, according to therapist's reports, extremely well adjusted to wearing a prostheses. Thus, their gait was very natural cosmetically. Figure 5.14 (top and bottom) illustrates the recordings of two patients. The deviation form normal gait may be observed, especially in the first subject. Furthermore, there are also noticeable differences between waveforms generated by the healthy and the prosthesis leg. In both patients, the small, anteriorly directed anteroposterior force peak of the prosthesis leg was lacking, while in signals by the healthy leg, as well as in Nilson's results of a healthy population this normally exists. Next, in the first patient, due to a compensation effect, the vertical force signal of the healthy leg deviated more from normal than in the injured leg. Data from this research were statistically processed and were potentially useful for the evaluation of the prosthesis adjustment procedure at the walking school.

Halett and colleagues[152] have researched the pathophysiology of posture and gait in individuals with cerebellar ataxia. They used the AMTI platform. They presented the signals of the vertical component of ground reaction force of a healthy subject and of a patient. The comparison revealed that there is an appreciable irregularity and weak formation of the initial phase of the force signal in the patient (Figure 5.15).

In the field of healthy locomotion, measurement data of normal running may be an important reference for analysis of running technique, the influence of athletic footwear, etc. Significant values of transversal force components can therefore indicate inefficient propulsion, while areas below the curves measured can point to energetic efficiency in walking.

Data on the path of the foot's center of pressure (in footwear) during support time in gait and running reveal interesting additional information. Besides vertical forces, Cavanagh and Lafortune[153] also researched the path of the center of pressure during support in long-distance runners. Results of the measured center of pressure paths (the Kistler platform) made by two groups of runners differing significantly in their techniques of contact are presented. Some realized initial contact with the rear, i.e., the mid-part of the foot (Figure 5.16). (The authors state two characteristic subgroups of subjects; however, there were no subjects who realized first contact with the forefoot.) First contact almost always occurs along the lateral border of the foot, but the center of pressure is either in the rear or in the mid-part of the foot. Individuals realizing initial contact with the mid-foot, land relatively flat-footedly and the force is more evenly distributed across the sole of the shoe. There is a weaker transition from the braking to the propulsion phase. Most of the contact-time center of pressure is between 60 and 80% of the length of shoe (measured from the heel). Implications of these measurements are interesting in connection with observing traumas caused by running—mid-foot strikers are exposed mostly to injuries and stress since the foot is not prepared for such loading.

There is a danger, however, of interpreting ground reaction force data without taking simultaneous kinematic information into account. Wrong conclusions may be

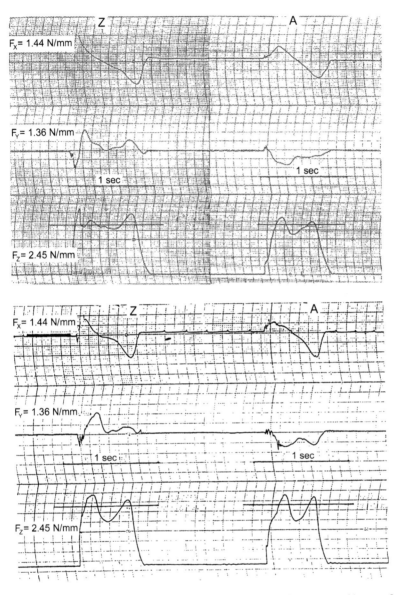

FIGURE 5.14 Ground reaction force signal waveforms in walking generated by two female patients (up and down) with an amputation below the knee, wearing a prosthesis. Z marks the healthy leg and A marks the leg equipped with a prosthesis. (Note the curvilinear plot.) (From Milinković, G., 1988. B.Ed. thesis. Comparison of Ground Reaction Force in Normal and Impaired Gait—Lower Leg Amputations, Zagreb, Croatia: Faculty of Physical Education, University of Zagreb. With permission.)

arrived at since, in the absence of kinematic information, there is no notion about corresponding moment of force in a certain joint (i.e., knee), which is, in effect, the decisive factor. Therefore, data of the measured force itself should be interpreted with reservation.

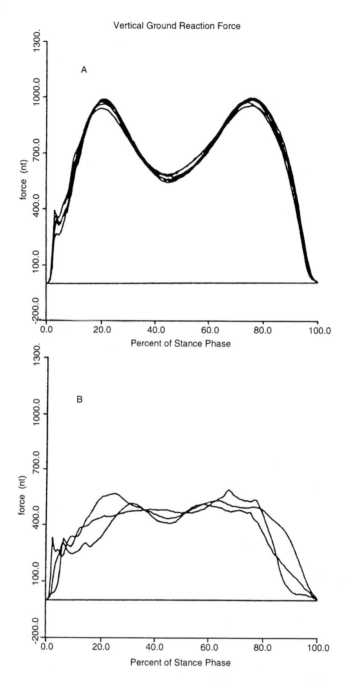

FIGURE 5.15   Ground reaction force in walking in a healthy person (A) vs. an individual with cerebellar ataxia (B). (From Hallet, M., Stanhope, S.J., Thomas, S.L., and Massaquai, S. 1991. *Neurobiological Basis of Human Locomotion*, M. Shimamura, S. Grillner, and V.R. Edgerton, Eds. Tokyo: Japan Scientific Societies Press, 275–283. With permission.)

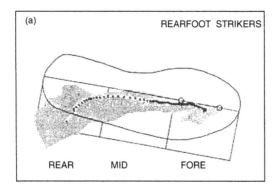

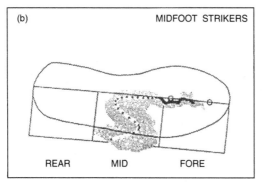

FIGURE 5.16   Mean center of pressure locations under the shoe outline at 2-ms intervals during contact. The shaded areas mark the range determined by normalizing all data on the basis of shoe length. Two subgroups of subjects with regard to the area of initial contact may be seen. (From Cavanagh, P.R. and Lafortune, M.A. 1980. *J. Biomech.* 13:397–406. With permission of Elsevier Science.)

5.1.4.2   Take-Off Ability

Testing jumping ability: The modified Sargent jump which includes use of the force platform (routinely applied in the Biomechanics Laboratory at the Faculty of Physical Education) will be described. Conceptualized according to the idea of Stjepan Heimer, the test evaluates the explosive power manifested at take-off by the lower extremities; it is simple and often used with athletes, typically volleyball and basketball players. (Explosive power denotes the ability to develop a large amount of mechanical work in one unit of time.)

At the start, the subject stands in a relatively upright position on the platform. He is required to perform take-off from both feet in a vertical direction with maximum intensity and to reach his hand up as high as possible on a laterally positioned (in a vertical direction) centimeter scale (jump and reach). Prior to take-off, knee bending and hand countermovement are allowed. Standard procedure includes three repetitions (trials). Besides the value of vertical reach, sagittal force component curves are also registered for each trial, F_x and F_z (Figure 5.17, top and bottom curves).[154] The best measurement trial is usually selected for analysis.

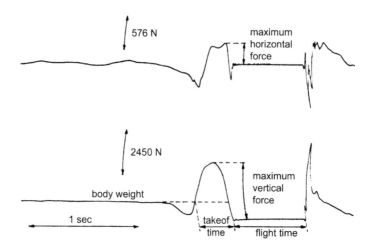

FIGURE 5.17 Anteroposterior (upper curve, F_x) and vertical (lower curve, F_z) takeoff force components in a modified Sargent test performed by a top-ranking basketball player. (Note the curvilinear plot.) The table shows complete measurement results in a highly selected sample of top basketball players. (From Milanović, D., Heimer, S., Medved, V., Mišigoj-Duraković, M., and Fattorini, I., 1989. *Basketball Med. Periodical* 4(1):3–8. With permission of the Croatian Olympic Committee.)

| examinees | measured kinematic data | | kinetic take-off data | | |
| | | | measured | | extracted |
	max. reach (cm)	max. reach from the spot (cm)	abs. max. take-off force (N)	act. take-off time (msec)	absolute max./body weight
V. S.	352	75	3580	220	3.1
K. T.	341	75	2685	240	3.1
D. V.	334	70	3246	250	2.9
P. Z.	321	56	2320	200	2.3
Z. J.	318	70	2762	200	3.2
C. Z.	314	60	2570	220	2.7
C. D.	310	59	2682	190	2.9
Ra. Z.	305	66	2608	170	3.2
R. Z.	306	69	2856	210	3.3

The value of the difference between jump reach and standing reach, expressed in centimeters, gives insight into the take-off ability of an individual. This feature, depending on the morphological and biomechanical-physiological characteristics of physical build (body height, length of lower extremities, etc.) and on the complex make-up of motor abilities, represents a very important factor of success in many sports disciplines, such as basketball, track and field, volleyball, etc. Top-ranking basketball players have produced values of around 60 to 70 cm, significantly surpassing values of the normal population.

The analysis of ground reaction force signals registered during take-off enables a deeper, albeit indirect, insight into the capability of developing explosive strength in the lower extremity skeletal musculature and with this take-off ability. The features of the curve $F_z(t)$ which describe impulsive waveform—maximum value of the vertical force component (absolute amount, also expressed relatively with respect to body weight in standing) and short duration (i.e., small width) and steep rise of a signal—are proportional to the explosiveness of contraction of the lower extremity extensor muscles (musculus rectus femoris, musculi vasti, m. gastrocnemius, m. soleus). Muscle explosiveness is by and large an inherent feature which can, however, also be influenced by directed exercise, i.e., training. Well known are plyometric exercises, specially aimed at the development of explosive strength. They are performed with a landing from a height of 40 cm followed by take-off. Therefore, if the Sargent jump produces a shape in the ground reaction force curve differing from the one described, this may be a guideline for the coach to try to improve the jumping ability in his athlete through adequate training. In this way, the reach will increase.

Based on experience with basketball players, some with top-ranking abilities, values for the force signal parameters have been determined: $\cong 3$ (ratio of maximum value of the vertical component of the take-off force curve and the amount of body weight when stationary), i.e., 200 ms (impulse width of vertical force), respectively, so that signal features of this kind may be accepted as an empirical criterion for good jumping ability.

The quantities mentioned were determined routinely from the results of a modified Sargent test, a practical and quick test. Further comparative analysis of force waveforms F_x and F_z, (especially if one also includes additional measurement quantities such as EMG signals) gives more profound insight and an estimation of the dynamics of the eccentric (during previous hand countermovement)/concentric contraction exchange of muscle groups responsible for vertical body take-off, a very important success factor in a number of sporting activities. This kind of information may help a coach to optimally use the reflex abilities of the tested athlete's neuromuscular system and the inherent elasticity of his muscles and tendons for a better performance result. Due to its noninvasiveness and the possibility of simple and fast application, the described test is a suitable diagnostic procedure for athletes of various disciplines. The following characteristics of a good measurement method are included in the test: repeatability, possibility of direct (on-line) data display, simplicity, and an affordable price. Having applied it, some experts state that it is extremely sensitive to the physical condition of the subject (Janković,[155] volleyball). It is usually applied as part of a broader test battery of medical, motor, and other tests, so that results may be interpreted simultaneously and presented to those interested (coaches) shortly after measurement. These procedures are therefore useful in diagnostics of the condition, follow-up, and direction of long-term training cycles in competitive sports.[154]

The test described is reminiscent of the old measurement method by Marey. With regard to performance of the jump itself, as well as registering the reaching-height achieved, the test differs from the classical one (according to Dudley Allen Sargent, physician and a well-known American innovator in the area of physical education at the beginning of the 20th century and founder of the Sargent Normal School of Physical Training, Cambridge, 1881; today this is the School of Allied Health Professions of Boston University[156]) in the following details: after the countermovement, take-off follows immediately, while in the classical variant, the subject is required to pause ("freeze") for some time in this position. So, modification of the test also encompasses the take-off force potentiation mechanism by means of a fast exchange of eccentric and concentric contraction of the extensor muscles of the lower extremities.

High jump take-off: Zmajić[157] measured take-off force in high jumps performed by the backward technique (the Fosbury flop) in laboratory conditions (Figure 5.18). The force signal waveforms recorded showed great individual specificity (Figure 5.19). The author interpreted such findings within the context of the performance technique for each individual subject.

Research into take-off ability of gymnasts: During a 2-year period, the take-off abilities of Zagreb gymnasts, ages 9 to 17 years, were researched. The sample initially encompassed 16 girls. Ground reaction force during take-off was measured while performing the following four technique elements: backward somersault and straddle jump, both performed from a standing position, and running forward somersault and layout forward roll, both preceded by a short run. Gymnastics judges simultaneously graded the quality of each performance.[158,159] To estimate the quality of performance in this manner is largely heuristic, including technical, esthetic, and other partially subjective factors, but standard procedure in gymnastics.

In backward somersaults from a standing position[160] (Figure 5.20), comparison of indicators deduced from take-off force signals and the corresponding performance grades pointed out that successful performances were characterized by signals of impulse shape and short duration, similar to those in the Sargent jump. (This refers to component F_z, while component F_x is significantly larger, in this case due to the initiation of backward body rotation needed to perform a somersault.) This is in agreement with the opinion of gymnastics experts who believe that a short and

FIGURE 5.18 Laboratory measurement of takeoff force in high jump, the Fosbury flop. (From Zmajić, H. 1988. B.Ed. thesis. Relations Between Biomechanical Parameters of Take-Off Force and Success in High Jump (in Croatian). Zagreb, Croatia: Faculty of Physical Education, University of Zagreb. With permission.)

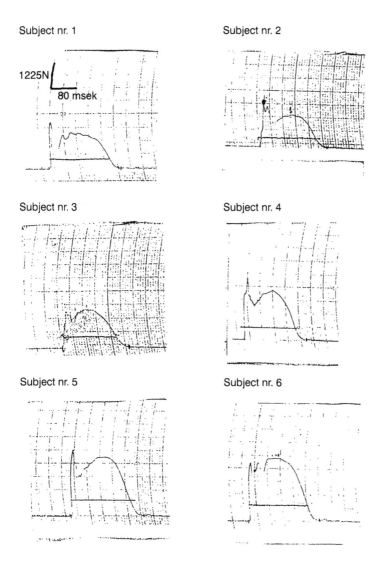

FIGURE 5.19 Vertical takeoff force component waveforms of several individuals performing high jump, the Fosbury flop. (Note the curvilinear plot.) (From Zmajić, H. 1988. B.Ed. thesis. Relations Between Biomechanical Parameters of Take-Off Force and Success in High Jump (in Croatian). Zagreb, Croatia: Faculty of Physical Education, University of Zagreb. With permission.)

energetic take-off is a necessary prerequisite for a successful performance, since it facilitates the performance of a high-quality somersault during the subsequent airborne phase.[161] A major observation in this research is the great individual take-off force waveform specificities during the concentration period prior to and during take-off itself.

Repeated measurements of the same subjects (9 out of 16) 14 months later revealed characteristic changes in the take-off force signal waveform, where some of

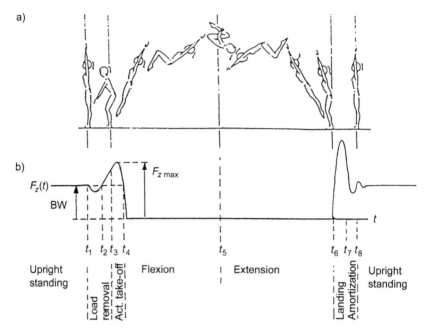

FIGURE 5.20 A kinogram representation of the backward somersault performed from a standing position (a) and a corresponding idealized waveform of the vertical ground reaction force component (b). (From Medved, V., Tonković, S., and Cifrek, M. 1995. *Med. Prog. Technol.* 21:77–84. With permission.)

the participants showed a tendency to "improve" the signal waveform shape toward a wave shape signal model of a successful take-off (Figure 5.21). Such changes could be explained by the fact that the subjects had acquired a motor stereotype for the backward somersault during this time period.

5.1.5 KINETIC SIGNAL REPRESENTATION

Examples of measurement signals presented have been in the form of time-dependent curves. There are now two additional ways of representation applicable in medical clinical practice and sports testing.

5.1.5.1 Vector Diagram

A vector diagram is a graphic representation of a spatiotemporal sequence of the two component vectors of ground reaction force in the sagittal, i.e., frontal, plane (F_x and F_z, i.e., F_y and F_z). This kind of representation might be provided after the signals are measured and A/D converted, if possible in real time, and is mostly realized in commercial systems with PC-supported measurement platform devices as a standard option.

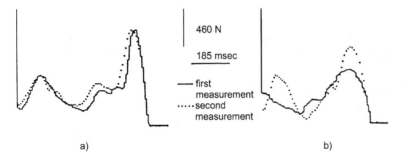

a)   b)

FIGURE 5.21 Takeoff force signals by two junior female sport gymnasts when performing backward somersault from a standing position, registered in a time span of 14 months. An appreciable increase in takeoff impulsiveness can be observed in the second gymnast; a sign of improved performance. (From Medved, V., Tonković, S., and Cifrek, M. 1995. *Med. Prog. Technol.* 21:77–84. With permission.)

Pedotti[162] and Crenna and Frigo[163] applied this kind of "synthetic" way to represent kinetic locomotion data in a clinical environment, resulting in a large number of gait measurement records. The features of signals so presented can be illustrated by taking the example of normal gait performed at three different velocities (Figure 5.22). The following characteristics of measurement records may be observed:

- Symmetry of records of left and right leg at certain gait velocity
- Harmonious and monotonous waveform of envelopes of curves
- Sensitivity of records to changes in speed of gait, in the sense of enlarging the difference between values of maximum and (local) minimum of the curve with increasing speed; shortage of duration of envelope and enlargement of the slope of vector in the beginning with respect to the end of the support phase
- Monotonous advancing of the point of the center of pressure in the direction of movement with a pronounced plateau during the last part of the support phase

At a certain speed, the records of a certain individual are repeatable.

Figure 5.23 shows several vector diagrams of gait in individuals with hereditary spastic paraparesy. Generally, individual deviations from the normal vector diagram model are present. The authors noted more inferior signal repeatability in these subjects than is observed in normal subjects, but among the signals shown, each one is nevertheless typical for the respective individual (steady state), and they are shown in the order of incidence of morphological changes of the envelope and, accordingly, to the degree of pathology. In this group of pathologies, the most frequent findings are as follows:

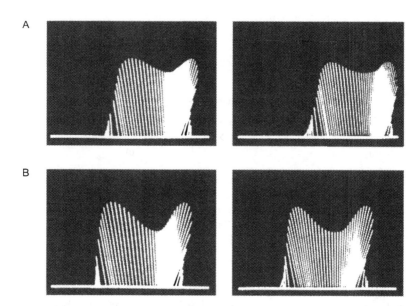

FIGURE 5.22 Vector diagrams of a normal subject at two different stride frequencies: 48 strides/min (A) and 56 strides/min (B). Each pair refers to the left and right foot (on the left and on the right side, respectively). Note the similarity between diagrams of the two sides, as well as changes to occur when walking speed is increased. Sampling time is 20 ms, progression is from left to right. (From Crenna, P. and Frigo, C. 1985. *Clinical Neurophysiology in Spasticity*, P.J. Delwaide and R.R. Young, Eds. Oxford: Elsevier. With permission of Paolo Crenna.)

- Increased vector amplitude in heel strike and considerable disorganization of the body weight acceptance phase
- Presence of a higher number of local maxima, resulting in a nonharmonious shape of the envelope and without smoothness
- General verticalization of vectors
- Inversion of the forward displacement of the point of application

The authors conclude that this kind of signal representation makes it possible to document and objectively follow a patient's recovery during rehabilitation.

These representations are well suited to man's heuristic mode of interpretation. They show sensitivity to particular locomotion abnormalities in the diseases/disturbances mentioned. Their interpretation, especially if done within the context of other (kinematic and kinetic, via inverse dynamic approach and/or EMG) possible data, may indicate specific lesions as causing the locomotion irregularity manifested.[162] Pedotti emphasizes the analogy between the possibility of interpreting such a representation of ground reaction force and the standard electrocardiographic (ECG) diagnostics of the heart muscle. It is further argued that the vector diagram feature substitutes multifactorial (ground reaction force, EMG, and kinematics) gait analysis.

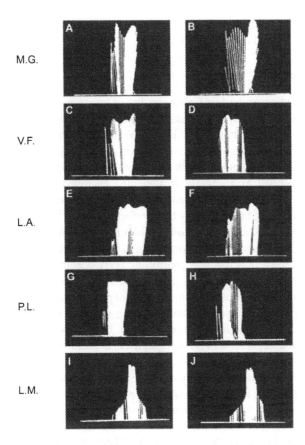

M.G.

V.F.

L.A.

P.L.

L.M.

FIGURE 5.23   Recorded vector diagrams from five individuals with hereditary spastic para-paresis with different levels of disability, listed (from top to bottom) according to the incidence of signal morphological changes. Barefoot walking without aids, at a natural speed 40 to 65 strides/min. Each pair corresponds to the left and right leg (at the left and right side, respectively). (From Crenna, P. and Frigo, C. 1985. *Clinical Neurophysiology in Spasticity,* P.J. Delwaide and R.R. Young, Eds. Oxford: Elsevier. With permission of Paolo Crenna.)

Santambrogio[164] developed an analytical method for a quantitative and detailed comparison of various ground reaction force patterns and suggested a pertinent index with which the differences estimated may be synthesized. The method is a statistical comparison of previously normalized ground reaction force signals generated by healthy and diseased individuals. The force amplitude is normalized by body weight value, and stationarity is also tested by calculating the time integral of the anteropos-terior force component which should equal zero (this denotes the existence of a neg-ligible inertial action in the direction of movement). The author claims that this kind of procedure enables reliable discrimination of healthy and pathological gait as well as the cases of minute differences which cannot be determined visually. Obviously, the next step should be clinical testing of the system by a diagnostics expert (coach or physician) and statistical evaluation of such objectivized diagnostic criteria, lead-ing to design of a clinical expert system.

5.1.5.2 Stabilometry

One typical application of the force platform device is to evaluate postural stability, i.e., body balance. In approximately static body postures, when an individual stands up straight on both legs, body stability is preserved if the vertical projection of the center of mass (CM) of the body falls on the foot support base. This area is deter-mined by "the lateral perimeter" of both feet. More precisely, the functional base of support may be defined as a slightly smaller area encircled by a border defined slightly within the support area. The reason for this is that muscles act against the whole body weight being taken over by the forefeet or heels.[165] In a healthy individ-ual, when the body is in a vertical position, it is never perfectly quiet and still, because spontaneous body oscillations occur (sway, which is the automatic action of neuro-muscular proprioceptive mechanisms controlling body posture). Through them, by compensating, a quasi-stable posture is retained. Visual, auditory, and vestibular sys-tems also act. In certain diseases or traumas, e.g., bad posture, neurologic balance dis-turbances, disability and the subsequently wearing of lower extremity prostheses, etc., characteristic deviations from normal occur in body position. The goal of sta-bilometry, a part of a larger field of posturography, is to quantify disturbances in pos-ture under these conditions.

In principle, to follow the movement of the center of body mass in space and time, the inverse dynamic approach should be applied, based on measured kinematic quantities and on biomechanical modeling of the body, while electromyography is needed to provide insight into neuromuscular function. These are complex proce-dures, however, which require the use of a comprehensive and expensive measure-ment system. A ground reaction force platform is, however, a device by which an approximate evaluation of time change of the position of body mass can be made using a simple, practical, and fast procedure. This leads to measuring (estimation of) body postural stability. A graphic display of measurement signals is usually used, providing an x-y diagram of the interrelationship between coordinate values of the center of pressure in the horizontal plane, determined by Expression 5.19. This out-put measure is a measure of body sway, albeit not the only one nor is it comprehen-sive. Based on dynamics principles of the system of particles, Shimba[166] developed a general, dynamic relationship between platform-generated data, the center of mass of a body on the platform, and the speed of change of angular momentum.

The actual testing procedures and protocols, as well as the processing of mea-sured ground reaction force signals, aimed at quantifying postural stability, are still not completely standardized. The validity of a number of procedures is being tested, and modifications of some classical neurological tests such as Romberg's are being made. Šimunjak's research[167] used recommendations from the 5th (Amsterdam 1979) and 6th (Kyoto 1991) International Symposia on Posturography which proposed the standardization of stabilometric measurements, the testing procedure, and the pro-cessing of signals measured in time and frequency domains. However, the fact remains that completely standardized criteria do not exist. Therefore, when

providing measurements, the procedure must be documented in detail. This is especially necessary since the majority of research reports show a high degree of inter-
and intra-subject variability concerning pathological findings.

In medical applications, stabilometry may be a part of the evaluation of the
action of particular drugs which influence the central nervous system. Hasan and co-
workers[165,168] found that psychiatric drugs (benzodiazepines) significantly increased
oscillations of the body. This research is primarily important in elderly individuals in
which there is a high risk of falls and fractures. Furthermore, these kinds of measurements may be applied to individuals with diverse cerebral deficiencies.
Overstressed responses detected in individuals with cerebral lesions may lead to
uncontrolled body movements of high amplitudes, even rhythmic oscillations. It
seems that part of the problem in such individuals is in the control of feedback from
the perturbation of posture. Hallett et all.[152] summarized abnormalities in stabilometric findings in connection with some cerebral diseases. Lesions of the frontal lobe are
characterized by large sway in the anterio-posterior direction with a 3-Hz frequency,
when the eyes are kept shut. Individuals with lesions in the loconodular part of the
brain manifest sway in all directions, with a frequency usually below 1 Hz, which is
pronounced when the eyes are closed (a Rombergisme). Lesions of the cerebral hemi-
spheres do not induce appreciable abnormalities in this test. Individuals with
Friedrich ataxia show increased lateral sway. Individuals with diffuse cerebral injury
also show abnormalities, but with a nonspecific cause.

With regard to the nature of neurophysiological mechanisms of postural and
motor control (Section 2.2), it may be rightly presumed that the approach to the study
and also the diagnostics of these kinds of quasi-periodical processes by means of frequency analysis might yield valuable knowledge. According to Collins and De Luca
(1993, according to Reference 169), there are two types of control—short-term open
loop and long-term closed loop—regulating body postural sway during a still upright
position. Recent research indicates nonstationarity of this phenomenon. Schumann et
al.[169] have carried out a time-frequency analysis of the curves of the center of pressure applying a nonstationary spectral estimation technique: the evolutionary spectrum according to Priestly. This kind of research has the perspective of identifying
parameters of the model of the subject-platform system and of further determining
and validating medically and kinesiologically applicable clinical criteria of postural
stability.

At the other end of the spectrum of applications of stabilometry are investigations of postural stability in top athletes. Heimer and colleagues[41] measured 13 top-
ranking female rifle shooters, to investigate the relationship between body stability
while aiming and shooting and the resulting score when shooting at a standard target,
under conditions similar to those normally encountered during training and competi-
tion. All participants used the same rifle. Postural stability is customarily evaluated
by determining the excursion of the point of center of pressure around the origin of
the coordinate system; the center of pressure is determined according to Expression
5.19. Since, however, the platform allowed measurement of only the force components, the values of the coordinates of the center of pressure were approximated using
values of horizontal ground reaction force components F_x and F_y. The attained maximum deflection values of the curves registered, F_x and F_y, during a 10-s time interval

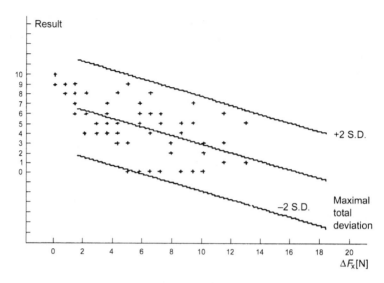

FIGURE 5.24 Relationship between the maximum amplitude values of horizontal ground reaction force curves in 12 top female shooters registered while aiming and shooting and the result and corresponding linear regression line. (From Heimer, S., Medved, V., and Špirelja, A. 1985. Proc. 2nd Scientific Congress in Sport Shooting—Medicine, Psychology, Pedagogy in Sport Shooting, Osijek, 47–56. With permission of the European Shooting Confederation.)

of aiming prior to shooting, were selected as a signal-based quantitative criterion. A significant inverse proportionality between the shooting score and the amplitudes of deflection of force components was found (Figure 5.24). Signals belonging to the most (Figure 5.25A) and least successful (Figure 5.35B) athlete reflect the span present in fine motorics in this demanding quasi-static physical activity in the measured subject sample.

A similar, but autonomous field is the development and application of mobile platforms, i.e., devices by means of which the human posture control system is subjected to mechanical disturbances, in an uncontrolled or controlled manner. This field of research developed during the last three decades, starting with Lewis M. Nashner's dissertation in 1970 at MIT and continuing with commercial devices, such as the expensive EquiTest in 1986 (according to Reference 170), a custom designed and experimentally manufactured movable platform,[171] or the interesting, new transportable device IBS (Interactive Balance System), manufactured by Tetrax and Neurodata, all enabling original and sophisticated clinical testing of balance and equilibrium. In spite of being very sensitive in some conditions, and often the methods of choice, certainly making them superior to stationary platforms, methods using these and similar movable platforms fall outside the scope of this book, since they concern perturbed postures. (This book is primarily directed at measurement of free, natural, and spontaneous as well as trained sporting locomotion. For the same reason, it does not encompass the various training devices or instruments which measure locomotor function and also provide the possibility of graded controlled mechanical loading of the organism.)

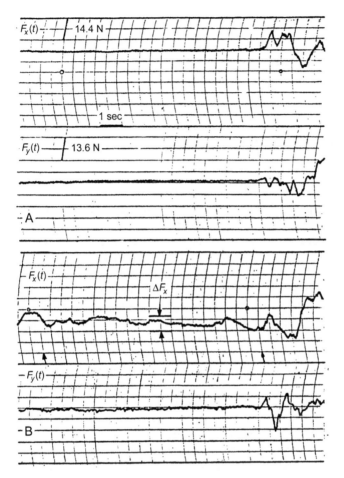

FIGURE 5.25   Horizontal components of ground reaction force signals while aiming and shooting, belonging to the best (A) and to the worst female shooter. (Note the curvilinear plot.) (From Heimer, S., Medved, V., and Špirelja, A. 1985. Proc. 2nd Scientific Congress in Sport Shooting—Medicine, Psychology, Pedagogy in Sport Shooting, Osijek, 47–56. With permission of the European Shooting Confederation.)

In summation, the basic advantages of the ground reaction force platform are

- Simplicity of measurement procedure and direct (on-line) result presentation capability
- Standardization
- High accuracy and precision, particularly in piezoelectric transducer-based platforms, but also in platforms using strain gage type transducers
- Possibility of application as a self-sufficient device, since it offers much more information than any one kinematic parameter (However, it is better to employ devices in combination with kinematic and/or myoelectric parameters.)

The basic limitations are

- Limited surface of the measuring area
- Problems with measurement invasiveness, i.e., "hitting" the platform
- High price, since these are highly precise and robust measurement instruments

5.2   PRESSURE DISTRIBUTION MEASUREMENT SYSTEMS

The kinetic measurement devices considered so far enable registration of instantaneous values of applied force and, possibly, the moment of force as resultant quantities which, hypothetically, act in only one point whose planar coordinates change in time. This is an idealized view, and the point of center of pressure may even be totally fictitious (i.e., fall outside the contact area), e.g., when an individual assumes a symmetrical two-legged, upward, standing posture. However, this reflects a view compatible with what was considered in Section 2.1, where the biomechanical model of the human body consisting of interlinked rigid segments was introduced.

In reality, body support occurs through a certain contact area between the foot or, alternatively, the sole and the ground, so that the total force of action is distributed. Therefore, distribution of pressure (defined as force over the unit area) over the ground must be considered. Existing technological solutions for measuring and registering pressure distribution between two rigid bodies offer new quantification possibilities for human biomechanics. By means of systems of this type (e.g., pedobarographs), mechanical interaction between the body, via the foot, and the ground may be followed in greater detail. In principle, there are three technical possibilities to detect pressure distribution.[13] The methods will be classified in three groups according to manner and site of application of the sensor unit on the subject, and not—also possible—according to the physical transduction principle, since it is the actual mode of application which determines the practical utility of the device. The technical possibilities are

- Install the sensor in the sole surface
- Install the sensor in footwear insoles
- Permanently embed the sensors in the ground, as is the case of a ground reaction force measurement platform (subject must be barefoot)

The historical development of corresponding technical solutions is briefly surveyed, based on the references.[13,23,172–175] A short description of recent solutions and the most important experimental findings are in Section 5.2.2.

5.2.1   HISTORICAL DEVELOPMENT

The first studies of dynamic contact between the foot and the ground date from Marey, Carlet, and Vierordt who developed methods based on pneumatic and similar principles (described in Section 1.2). It is obvious that devices of this type were not, according to standards today, precise or noninvasive enough. In their time, these methods were an expression of the natural effort to measure locomotion in a better way and they may be considered precursors to procedures for registering body-ground contacts and measuring ground reaction forces in general. However, the real precursors to pressure distribution measurements, with planar resolution of an order of magnitude cm × cm required, appeared (coincidentally to the technological development of materials) relatively late, only in the mid-1970s. In the course of this development, as today, there are many interesting engineering solutions, only a part of which presented.

Development of footwear with embedded sensors: Development in this field initially began with Marey and Carlet's methods. Only in the early 1970s were efforts made to locate transducers on or in the sole. Spolek and Lippert used strain gage-based spring elements, positioned under the forefoot and the heel, to measure vertical and horizontal components of reaction force. An axial moment could also be measured. However, the device was rather clumsy, albeit not invasive, according to the authors. Miyazaki and Iwakura used a strain gage mounted on a metal plate and fixed to the shoe sole to measure vertical force. Drawbacks to the device were variable transducer sensitivity with various forms of foot and shoe hardness and invasiveness. Miyazaki and Ishida developed another procedure, with two large capacitive transducers per shoe, but it was possible to measure only total force.

Footwear equipped with transducers was described by Ranu in 1986. He measured ground reaction force by installing two 8-mm-thick three-axial load cells in the sole. Several strides in succession and displacements of the center of pressure could be measured. However, considerable modification of the shoe was required which is not clinically practical.

Development of insole-based systems and platforms: The first efforts to register foot imprints on deformable materials such as plaster of Paris and clay date from 1882 (Beely, according to Reference 173) and were described by Abramson (1927, according to Reference 174). These techniques did not allow real measurements, however, and the resulting information related in a large measure to the form of the foot rather than to pressure distribution. Morton (1930, according to References 173 and 175) was the first to apply the electrical properties of rubber. His "kinetograph" gave a foot imprint on inked paper by means of a special rubber mat. The upper surface of the mat was flat, while the lower one was ridged and coated with ink. Localized pressure on the upper surface flattened the pyramid-shaped bulges in the direction toward the ground and left an ink imprint consisting of parallel lines whose width was proportional to the pressure. This resulted in good-resolution images. Various optical and electromechanical techniques were applied. By using a similar deformable mat, Elftman (1934) analyzed gait dynamically by film recording the change of distribution of plantar pressure through a glass plate. At the same time, Schwartz conducted his pioneering studies (Section 1.2), in which he identified and quantified parameters characterizing normal gait through an analysis of plantar bearing. The rubber elasticity principle was also used by Harris and Beath (the Harris and Beath footprinting pad). The next modification was the replacement of the paper ink record with thin aluminum foil (Grieve). Through this, a permanent imprint of applied pressure remained on a foil which could be interpreted visually, but also quantified by optical scanning (foil pedobarogram). These were rather inexpensive methods.

It was not until 1941, however, that suitable transducers for registering time changes in pressure at various locations under the foot were developed. A number of methods were proposed for the visualization of a 2D pressure map between the foot and the ground (Hertzberg 1955, Scranton and McMaster 1976, according to Reference 174). Scranton and McMaster employed a liquid crystal display in the aim of visualizing and qualitatively estimating pressure distribution under the foot in normal and ill individuals. However, neither the frequency characteristics nor the input-output characteristics (deviating from linearity) of these methods were satisfactory. Strain gage silicon sensors were also used (Pratt et al. 1979, Lereim and Serk-Hanssen 1973, according to Reference 174) in which the pressure between foot and insole, i.e., insole and ground, was detected, both in static posture and while walking.

Arcan and Brull (1976, according to Reference 175) constructed a device in which foot pressure was transmitted onto a photoelastic mat by means of an array of 500 semicircular buttons. Pressure distribution below the foot was visualized, and loadings at each point were analyzed as a function of the diameters of obtained interference rings. By using this device, Simkin and Stokes (1982) developed computer programs for the analysis of plantar dynamics and the representation of force distribution below the foot. Afterward, Betts et al. (1980) studied changes in pressure which were reflected by light intensity reflected on a plastic/film interface and developed automated computer based image processing methods (Duckworth et al. 1982).

A solution was developed (Nicol and Hennig 1978, according to Reference 174) based on a parallel-plate type of electrical capacitor, the capacitance of which was a function of pressure. Capacitors (256 maximum) were filled with foam rubber and arranged into a flexible matrix, forming a pressure mat. This was the only method suitable for direct analog readout of pressures at discrete locations under the foot. The mat consisted of an upper layer of conductive plates connected in rows and a lower layer of conductive plates connected in columns, separated by an elastic rubber material, a compressible dielectric. The dimension of the capacitors was 2.1×2.3 cm, and the total size of the mat was 20×40 cm. By applying pressure to each individual transducer, its capacity increased and its impedance decreased. The authors indicated limits to their technique: low resolution due to rather large transducers, low correlation between foot contour and transducer location, limited 25-Hz sampling frequency, and limited precision compared to conventional transducers, but the advantages of being inexpensive and portable. The system was a predecessor to the commercial system EMED.

At The Pennsylvania State University, insoles equipped with 400 and 1000 sensors (ferroelectric polycrystalline ceramic transducers immersed into silicone rubber) were designed, constructed, and used. While the limitations of the previously mentioned capacitive sensors were due to their nonlinear stress-strain characteristics and the hysteresis phenomenon, the basic drawbacks of ferroelectric ceramics impregnated by rubber were their relative fragility, low resistance to fatigue, and complicated construction. Measurement signals were fed into a computer and a graphic display via plotter was used. Figure 5.26 illustrates this impressive way of representing measurement information.[176] Normal gait was measured at the time instants indicated. During the support period, a gradual transition of the center of pressure from the heel toward the forefoot along a characteristic trajectory may be observed. These results broaden information on the measurement of the trajectory of the center of pressure (Section 5.1.4).[153]

Pedotti et al.[174] described a technical solution for a shoe insole equipped with piezoelectric sensors with PVDF (polyvinyl fluoride). This piezoelectric polymer incorporates the good features of polymer materials: light weight, flexible, and easy-to-process technology with a high piezoelectric coefficient, a broad frequency band, and linear output in a broad dynamic range. The system had 16 sensors arranged on a foot, whose outputs were connected to a computer by means of a 16-channel A/D converter with a multiplexer. Time resolution thus achieved was 100 Hz per channel. Signal processing encompassed normalization of absolute values of particular sensor outputs with regard to the largest value measured (and not to body weight; a value which is surpassed in walking) which the authors advocated as clinically being the most relevant.

5.2.2 SOLUTIONS AND EXAMPLES OF USE

Today there are a number of instruments on the market and in laboratories for measuring pressure distribution between the foot and the ground which can be applied in the study of posture and locomotion. Besides problems occuring in sports medicine, physiatry, and orthopedics, syndromes (pathologies) from other medical fields can also be evaluated indirectly by means of these devices. In diabetes, for instance, anomalies in circulation develop and are reflected particularly in the foot. Pressure distribution data may therefore offer new and original information important for treatment. Such measurements may provide a basis for manufacturing insoles aimed at correcting irregular pressure distribution and preventing pathological effects.

Solutions presented are those most applied, according to the references. A great majority of them only enable registering of the vertical pressure component, but there are solutions for measuring horizontal components as well. The classification according to Cobb and Claremont[172] is followed.

5.2.2.1 Platforms

Since the early 1980s, several instruments for measuring pressure distribution in barefoot walking and standing have been developed.

Musgrave Footprint System: The Musgrave Footprint System (by Musgrave Footprint, Preston Communications Ltd., Llangollen, Clwyd, Great Britain) uses resistive force sensors. The sensor consists of two thin sheets of polymer material, one with an embedded comb-shaped electrode (Figure 5.27) and another coated with the semiconductor material, molybdenum disulfide. The physical transduction

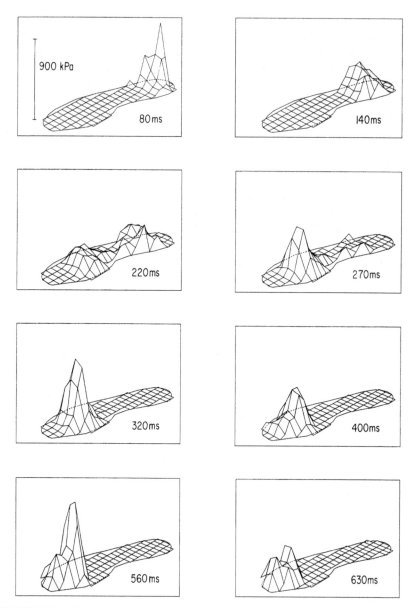

FIGURE 5.26 Computerized 3D display of pressure distribution during normal walking in shoes. Selected displays obtained from the support phase in normal slow walking in shoes. The height of the surface above the ground is proportional to the pressure between the shoe and the ground. A calibration bar for all surfaces is shown in the 80-ms display, and times are shown in milliseconds after first ground contact. Hidden lines have been removed except for shoe outline which has been dotted where it is obscured by the pressure surface. (From Cavanagh, P.R. and Ae, M. 1980. *J. Biomech.* 13:69–75. With permission from Elsevier Science.)

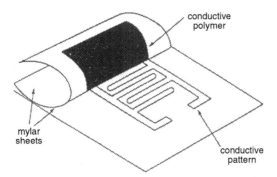

conductive
polymer

mylar
sheets

conductive
pattern

FIGURE 5.27 Structure of the force-sensing resistor characterized by Maalej et al. (1988), applied for measuring pressure distribution in the Musgrave system. (From Cobb, J. and Claremont, D.J. 1995. *Med. Biol. Eng. Comput.* 33:525–532. With permission.)

principle relies on the premise that the contact surface between electrodes and semi-conductor material increases with applied force, resulting in a large change in resistance. The thickness of such systems ranges from 0.25 to 0.7 mm. A transfer characteristic is logarithmic, where the precise response is a function of the type of surface, conductor geometry, and semiconductor material used. The system includes a matrix of 2048 resistive sensors, 3×3 mm in dimension, with a measurement range from 0 to 4 MPa per sensor. The device may be used in a range from 11 to 110 kPa, giving a typical change in resistance from 1 $M\Omega$ to 2 $k\Omega$. The sensor's response to a given load may vary by $\pm$ 15%. The advantages of the instrument are that it gives high-resolution imaging and may be well incorporated within the walking path, while the drawbacks are inherent problems with resistors, low repeatability, and nonlinear sensitivity.

Pedobarograph: In the pedobarograph (Betts et al. 1980, Franks at al. 1983, according to Reference 172), the upper surface of a glass plate is coated with a thin nontransparent material, usually a thin plastic sheet. With applied loading, changes in the level of contact at the glass/plastic interface result in a change of refraction index and attenuation of light spreading through the glass plate. When viewed from below, contact surfaces (between the pressed layer and the upper plate) are seen as low intensity regions. Figure 5.28 depicts a pedobarograph and behavior at the glass/plastic interface. The system's quality depends on video camera quality, gray scale conversion, and plastic sheet characteristics. These systems do not allow focal calibration since there is no electrical access to individual sensors. Calibration is also rather difficult. By carefully selecting a transducer sheet material, an approximately linear relationship between applied pressure and light intensity can be achieved. In spite of problems in clinical application, the system has gained great popularity.

System by Rhodes et al.: Rhodes et al. developed a high-resolution imaging technique by using a photoelastic polyurethane sheet $500 \times 380 \times 2.4$ mm in dimension and a polarizer, fixed to a walking path constructed of 19-mm-thick transparent acryl. A thin sheet of silver-painted polycarbonate above the photoelastic plate acts

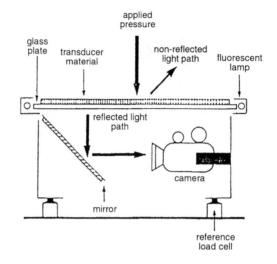

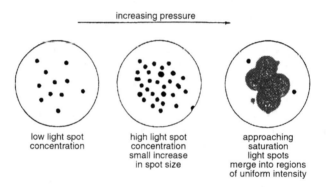

FIGURE 5.28 Elements of the pedobarograph transducer system and behavior at the plastic/glass interface. (From Cobb, J. and Claremont, D.J. 1995. *Med. Biol. Eng. Comput.* 33:525–532. With permission.)

as a reflector. Loading is transferred to the photoelastic plate through a 3-mm-pitch corrugated plastic indenter to realize a maximum difference in two orthogonal directions of strain, to which the photoelastic effect is proportional. By increasing the loading, a light transmitted through the photoelastic plate rotates around the propagation axis and may pass through the polarizer to form the parallel line image of an indenter, with point intensity proportional to the applied load. A picture is taken using a camera located next to the walking path and each pixel has a unique area of 0.96×0.79 mm. To filter out surface irregularities and simplify representation, the average intensity value in each 3 mm^2 area is attributed to the pixels. Calibration indicates nonlinearity errors smaller than 3%.

System by Assente: Assente et al. constructed a platform with 580×380 elements by using poled 40-μm piezoelectric PVDF (polyvinylidene fluoride) film, bonded onto a double-sided printed circuit board containing 7 × 10 mm rectangular copper pads. Each of the 1024 pads is capacitively coupled onto piezoelectric film and connected to instruments on the rear side of the board with true hole plating. The upper film side is pre-platinated with aluminum to realize common reference, the mass. Charge amplifiers sample the output of each transducer using a 100-Hz frequency.

5.2.2.2 Shoe Insoles

Discrete Sensors

System by Soames et al.: Soames et al. developed a transducer based on a variable resistance principle. It is 0.9 mm thick by 13 mm^2 and is made of beryllium copper. The pressure-sensitive area is a 3 mm by × 5 mm strain-gaged cantilever beam which, when loaded, deflects into a 0.4-mm recess milled into the underside of the transducer (Figure 5.29). To obtain accurate readings, the soft tissue must be compliant enough to distort the beam and the transducer must rest on a relatively rigid surface to avoid obstruction of recess on the undersurface. A rigid undersurface also secures even loading of the transducer. Calibration curves produced by hydraulically distending a rubber membrane over the transducer showed a linear relationship between applied force and voltage output. Besides problems of adequate in-shoe positioning, these low-dimension transducers (up to 15 per foot) are repeatable and clinical results are excellent.

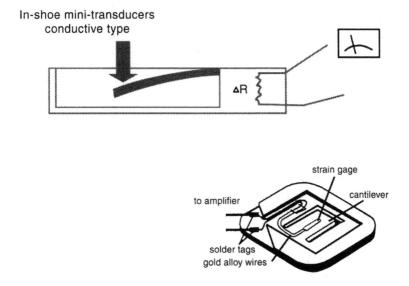

FIGURE 5.29 Conductive type transducer by Soames et al. (From Alexander, I.J., Chao, E.V.S., and Johnson, K.A. 1990. *Foot Ankle* 11:152–167. With permission.)

Matrix Sensors

EMED system: The EMED (Elektronisches Meßsystem zur Erfassung von Druckverteilungen) system is a conceptual continuation of the mentioned solution by Henning and Nicol based on capacitive mats and individual sensors. The system is made in the form of a platform and insoles. It has 99 sensors in an insole of 2-mm thickness and each individually occupies an area of 17 mm^2. Today, the solution offered is a multicomponent system capable of sampling 150,000 sensors per second. This number is a final determinant of resolution of the size of the active area and of the device's sampling frequency. The EMED SF system (Novel, Inc., Munich, Germany) has 2 sensors per square centimeter, an active measurement area of 27.4 cm $\times$ 48.8 cm (platform), and a speed of 70 images per second. Each sensor has an individual calibration curve, and the accuracy of the value presented is satisfactory. Also, 2-mm-thick insoles can be used. There is good software support, enabling particular foot regions to be chosen for measurement.

F-Scan system: The F-scan system is a commercial product by the U.S. firm Tekscan, Inc. (Boston, Massachusetts). It is based on a very thin, flexible, resistive tactile sensor, developed originally for measurements of dental occlusion, whose manufacturing methodology was originally developed for flexible printed circuits. It houses 960 sensor sites at the surface, each capable of 8-bit pressure resolution. The F-Scan sensor is shown in Figure 5.30. It is based on a combination of conductive, dielectric, and resistive inks. The sensor is characterized by a grid of rows and columns formed of silver-based conductive ink deposition. Each sensitive trace is coated with pressure-sensitive resistive ink, so that one sensor cell is created on each grid crossing point. The resistance of each sensor cell is inversely proportional to the applied surface pressure. By scanning. the grid and measuring the resistance at each crossing point, pressure distribution at the sensor surface can be determined. A unique feature of the manufacturing process of the F-Scan sensor is that the layouts may be adjusted to a broad spectrum of shoe sizes: the multilayer printing technology enables connection to traces forming the sensor grid at locations intermediate to their endpoints. A flexible equivalent of a multilayer circuit board is created by printing isolation dielectric coating across traces which connect the sensor with scanning electronic circuits. The small approach holes enable connection to the sensing area traces. Before depositing the grid traces, holes are filled by conductive ink to form a flexible equivalent of a plated-through hole. In this manner, the grid trace endpoints may be trimmed to contour sensor outline, while total functionality of the remaining sensor surface is kept. Electronic circuits for sensor signal measurement have two components. The ankle pack (cigarette box size) controls actual grid scanning and A/D conversion of measured resistance values. Digital data are transferred serially by a light coaxial cable into the remaining part of the electronics, located on a 3/4 length expansion card in a PC (VGA graphic support, at least 40 Mb). The system is computer-supported so measurement data may be presented in real time or stored and presented later in a number of detailed graphic modes. Among the first applications of the F-scan system was the evaluation of dynamic plantar pressure during walking in

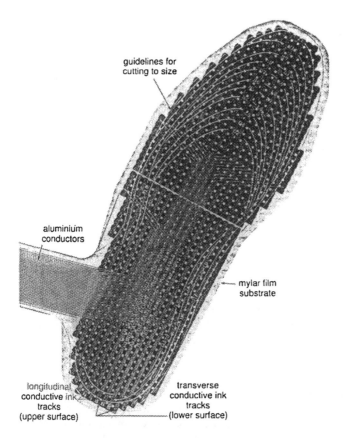

guidelines for
cutting to size

aluminium
conductors

mylar film
substrate

longitudinal
conductive ink
tracks
(upper surface)

transverse
conductive ink
tracks
(lower surface)

FIGURE 5.30   Matrix type Tekscan F-scan sensor in the form of an insole. Each resistive ink sensing element is formed between overlapping transverse and longitudinal conductors. (From Cobb, J. and Claremont, D.J. 1995. *Med. Biol. Eng. Comput.* 33:525–532. With permission.)

an individual with a high degree of metatarzaglia who had difficulty finding adequate footwear.[177,178]

Other Systems

Systems for measurement of tangential loading: Systems for the measurement of pressure distribution presented so far consider only the vertical pressure component (except Ranu). Tappin and Pollard (1980, according to Reference 172) developed a discrete transducer of tangential (shear) force component and the same magnetoresistive principle was used in the design by Lord et al.[179] (Figure 5.31). Relative movement of a permanent magnet with respect to a magnetoresisitive element results in a change of resistance in proportion to applied force. The elasticity of silicone rubber connecting two halves of the transducer gives a restoring force for the return of the magnet to a state of equilibrium. The maximum excursion of 0.6 mm corresponds to tangential (shear) loading of 250 kPa. By aligning a locking groove,

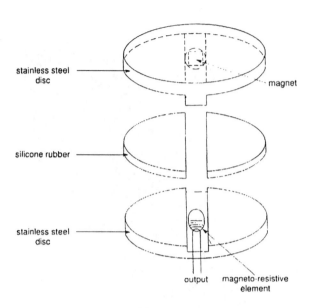

stainless steel
disc

magnet

silicone rubber

stainless steel
disc

output   magneto-resistive
element

FIGURE 5.31 Elements of the shear transducer designed by Lord at al. (1992) based on the magnetoresistive principle according to Tappin. (From Cobb, J. and Claremont, D.J. 1995. *Med. Biol. Eng. Comput.* 33:525–532. With permission.)

either in a longitudinal or a transversal direction with respect to the foot, the related shear component can be measured. The bridge circuit gives a temperature-compensated output voltage. Frequency response of the system exceeds 500 Hz, while clinically relevant are components up to 70 Hz. Precise transducer positioning is critical for application.

5.2.3 CLINICAL FINDINGS AND STANDARDIZATION OF MEASUREMENT

A summary of experimental findings on normal foot pressure distribution during walking was given by Alexander et al.[13] In barefoot walking, heel strike occurs initially with the posterolateral aspect of the heel and peak heel pressures are not reached until approximately 25% of the stance phase at which point the heel, lateral midfoot, and metatarsals all make contact with the ground. The velocity of the center of pressure is very high in initial heel strike, marking a rapid forward transfer of force. After heel strike, the velocity of the center of pressure decreases initially, but then accelerates again to rapidly cross the midfoot. Midfoot pressures are usually low. By transferring weight from the hindfoot to the forefoot, the center of pressure passes through the midfoot region, however, representing an average of forefoot and hind foot forces rather than true peak pressure in the midfoot area. The center of pressure is located in the forefoot for up to 40% of the time, although there are individual variations.

Soames et al. (1985, according to Reference 13) extensively researched the influence of footwear on pressure distribution using the previously described system (ten sensor sites). Footwear reduced peak apex pressure, producing a more even distribution of load under the heel (Figure 5.32). Forefoot effects were also significant. Weight bearing at a metatarsal head level shifted from the central metatarsals barefoot to the medial side of the forefoot in shoes, with maximum pressures under the first and second metatarsals (Figure 5.33). In shoes, contact times of the toes ranged from 60 to 85% of stance, a considerable increase from 50 to 55% of stance for barefoot walking. Peak pressures under the toes in shoes were also generally higher (Figure 5.34). The resulting higher foot-floor impulse under the toes in shoes was associated with reduced weight transfer by the metatarsal heads. There is a tendency for high-heeled shoes to load the medial metatarsals while unloading the lateral metatarsals. By using conventional methods, others have also found a pronounced tendency for this kind of footwear to reduce forefoot impulse and shift weight bearing to the midfoot and hindfoot.

In individuals with diabetic neuropathy, excessive local pressure correlates with foot ulceration (Cavanagh et al. 1985, according to Reference 172). These abnormally high pressures result from not achieving optimal pressure distribution, a consequence of impaired foot sensitivity. By measuring force distribution, one may therefore enable preventive measures to be applied, diminishing risks of further complications. This is the field in which pressure distribution measurements are clinically most extensively applied. To indicate areas of high risk, dynamic measures have

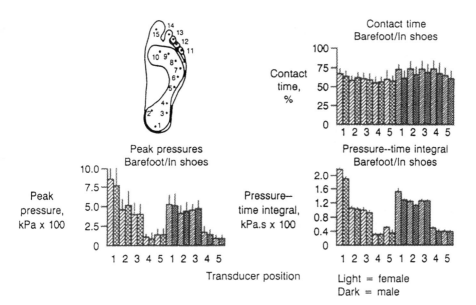

FIGURE 5.32 Transducer measured plantar pressures at hindfoot and midfoot during normal walking. Calculated from Soames. (From Alexander, I.J., Chao, E.V.S., and Johnson, K.A. 1990. *Foot Ankle* 11:152–167. With permission.)

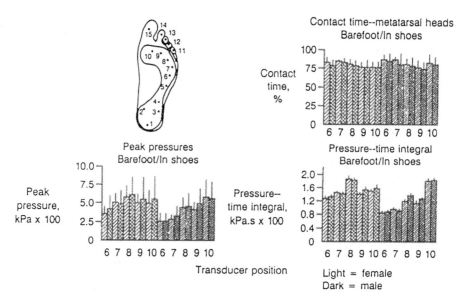

FIGURE 5.33 Transducer measured plantar pressures at forefoot during normal walking. Calculated from Soames. (From Alexander, I.J., Chao, E.V.S., and Johnson, K.A. 1990. *Foot Ankle* 11:152–167. With permission.)

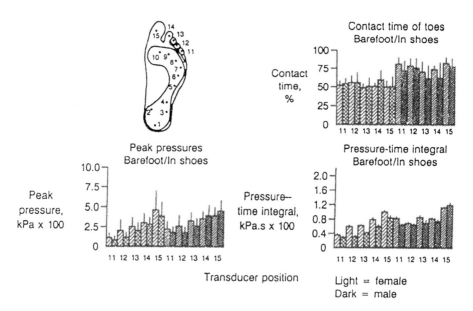

FIGURE 5.34 Transducer measured plantar pressures at toes during normal walking. Calculated from Soames. (From Alexander, I.J., Chao, E.V.S., and Johnson, K.A. 1990. *Foot Ankle* 11:152–167. With permission.)

proven more effective than static. Apart from vertical forces, ulceration areas also correlate well with areas of maximum horizontal forces, valid for individuals with healed and active ulceration. The progression of neuropathic changes relates to abnormality in increased foot loading. In neuropathy, weight bearing shifts from the medial to the lateral border of the forefoot.

A rheumatic foot also resembles a diabetic one. Toe loading is decreased, while lateral metatarsal heads are increasingly loaded (an unexplained finding). Another clinical application of measuring plantar pressure distribution is by monitoring the effects of degenerative changes to the foot, such as in leprosy (Patil and Srinath 1990, according to Reference 172) and in postoperative evaluation after corrective surgical treatment in conditions such as hallux rigidus (Betts et al. 1980, according to Reference 172).

Another important application of this sort of measurement is in the design and manufacturing of athletic footwear. By measuring plantar pressure distribution and combining this information with other relevant information, individualized manufacture of customized shoe insoles can be accomplished, with possible broad application in the fields of sports recreation and sports medicine. In the Peharec Center for Physical Medicine and Applied Kinesiology in Pula, Croatia, the F-Scan pedobarographic system is used with the goal of recording the characteristics of foot imprints while standing, walking, and running, from which customized insoles will be manufactured. A number of athletes have been measured at the Peharec Center, especially top-ranking handball and basketball players. Insoles should increase the foot support area and thereby diminish pressure in regions of maximum values where ulceration risk is maximum. This is applicable both in athletes and in normal subjects.

Shear transducers were applied in clinical tests: the influence of four different types of footwear was studied in ten normal subjects (Pollard et al. 1983). Tappin and Robertson measured barefoot (10 subjects) and in-shoe plantar pressures (20 subjects). Results indicated that the amount of vertical force necessary to induce occlusion of blood flow decreased to 50% when shear force is simultaneously present.

While there is no doubt about the relevance of clinical and other diagnostic applications of the described pressure distribution measurement systems, standards are still developing. Suitable clinical protocols to be applied in the fields of orthopedics, physical medicine, and rehabilitation and sports medicine, which would qualitatively suit and supplement the other indices obtained by examination, are also being developed. As early as 1991, at an EMED meeting held in Vienna, Ian Alexander suggested a standard form for reporting foot measurements. Of the measurement devices treated in this book, these devices, together with those of 3D kinetics as developed from inverse dynamics, are the least standardized. Cobb and Claremont[172] concluded that the future development of plantar pressure transducers may include insoles capable of simultaneously registering vertical and horizontal pressure. Improved calibration methods are needed to make estimation of factors such as bending force and temperature on transducer performance possible. Standardized methods of evaluation and performance definition of plantar pressure transducers still need to be developed.

Advantages of systems measuring pressure distribution are

- They offer spatially precise information, new and original. Redundancy inherent in this information is still to be determined.
- Insole layouts enable measurement of more strides, giving them an important advantage over imbedded platforms because insight into statistical features of more successive strides is possible, important in population studies or in sporting contests.

Disadvantages of the systems are

- Precision and accuracy of measurements are inferior to ground reaction force platforms with piezoelectric sensors or strain gages and pressure measurement insoles wear quickly with use.
- Platform layouts measure only one step (a disadvantage of classical platforms as well).

6 Measurement of Myoelectric Variables

Kinetic and kinematic measurements provide either direct or indirect insight into the values of the resulting forces and the moments of force acting between the body and its surroundings, i.e., at the imaginary connections of particular body segments. In addition, a subgroup of kinetic measurement methods provides pressure distributions at the supporting surfaces. A more-detailed and fundamental physical study of the behavior of various biological tissues under the influence of mechanical loading caused by movement, or by static loading, is outside the scope of this book. The reader may refer to the literature.[2,24,180,181] Furthermore, by measuring kinematic quantities only and measuring and/or estimating kinetic (dynamic) quantities, the energy balance of movements measured may be estimated.[12,182-185] While this falls outside the scope of this book, it should be mentioned that in the biomechanics and physiology of activity, the "gold standard" and criterion for energy expenditure of an organism are procedures with the Douglas bag, a treadmill, and pulse and oxygen consumption measurements. These procedures should therefore, when possible, be applied simultaneously with the biomechanical measurements of the locomotions concerned.[1,186]

The next step toward profound locomotion analysis is based on a more-detailed modeling of the anatomical structure of the locomotor apparatus and, specifically, efforts to estimate contributions of particular muscular groups to the total resultant forces and moments in the locomotor apparatus. Attention is directed to the skeletal muscle viewed as a force actuator and a neuromechanical converter, characterized at the macroscopic level by the features of elasticity, viscosity, and mass (Section 2.3). Because muscular force is not practically susceptible to measurement *in vivo* noninvasively, electrical processes in muscles are accessible for measurement by means of the electromyography method. This method concerns the measurement (detection and amplification) and registration of changes in the motor unit's electrical action potentials accompanying the process of muscular contraction (Section 2.3). Bioelectric muscle activity may provide insight into bioinformation and biocommunication (muscular coordination, speed of muscle activation) and the bioenergetic aspects of muscle function (energy expenditure necessary for muscular tissue metabolism, local muscle fatigue), both of which are of interest to the study of locomotion, healthy and pathological.

This chapter begins with a historical survey of the emergence of the connection between the neuromuscular system and bioelectricity (Section 6.1). The model of the process of genesis of myoelectric changes is described next (Section 6.2). Under kinesiological electromyography, a presentation of the technique of classical surface electromyography, EMG telemetry, and myoelectric signal processing applicable in the study of human locomotion is given. Finally, results of electromyographic measurements of normal and some sports locomotion are presented and discussed (Section 6.3).

6.1 THE NEUROMUSCULAR SYSTEM AND BIOELECTRICITY: A HISTORICAL SURVEY

In Chapter 1, Luigi Galvani (1737–1798) was mentioned. This Italian Renaissance physicist and physician researched the bioelectricity of frog muscles. Here, further historical events important for an understanding of bioelectricity as a phenomenon are cited based on the references[32,51,149,187–189] as well as a paper by Clarys[190] introducing some new and contradictory historical notions.

Herzog et al.[189] referred to the work by Croone (1664, "De ratione motus muscularum") in which, based on the experiments on neural sections, he concluded that the brain must send signals to a muscle for it to contract.

The first to recognize the connection between muscles and the generation of electricity was Francesco Redi (1626–1698) in 1666. He was an Italian physician, natural scientist, and poet. He hypothesized that electrical shock in electric fish originated in the muscle (Biederman 1898, according to Basmajian and DeLuca[51]). Galvani, however, was the first (in 1791) to observe the connection between electricity and muscular contraction on a preparation of frog m. gastrocnemius (taken as the beginning of neurophysiology, "De viribus electricitatis in motu musculari," 1792). He showed that electrical stimulation of muscular tissue produces contraction and force.[189] About this time, an Italian physicist named Alessandro Volta (1745–1827) was also active. It is to him that we owe the invention of the galvanic cell. This technical possibility to generate electricity in some ways overshadowed Galvani's findings. Because there was no adequate, sensitive detection instrument available, 40 years passed before Galvani's discovery was finally approved and accepted. Alexander von Humboldt (1769–1859), a German naturalist, designed primitive electrodes and experimented with many forms of stimulation.

Based on a survey of original materials from university libraries in Heidelberg, Montpellier, Leiden, and the Belgian National Library, Clarys[190] presented data which somewhat changed the historical picture of the beginnings of electrophysiology. According to Boerhaave et al. (1737, according to Reference 190), Jan Swammerdam (1637–1680), a Dutch zoologist (anatomist and biologist), had already described various experiments on stimulation, depolarization, and contraction of nerves (i.e., muscles) in 1658. The experiments were basically very similar to those described by Galvani and von Humboldt, but were conducted 134 years earlier. Clarys elaborated the contents of written material accessible to him, hypothesizing on the lack of mutual communication among scientists of that time, in an attempt to explain certain omissions in the historical chronology of electrophysiology as we know it.

The first practical galvanometer by Johannes S.C. Sweigger dates from 1820 and is based on Oersted's discovery of magnetism. Nobili increased the sensitivity of this instrument by compensating for the influence of the Earth's magnetic moment. He measured the injury current by connecting his so-called astatic galvanometer to electrodes located on a cut surface and an intact surface of a frog muscle. Finally, in 1838 Carlo Matteucci proved that bioelectricity was connected with muscular contraction by using such an instrument.[51] In 1842 he demonstrated the existence of the action potential accompanying the contraction on the muscle of a frog.[149] He conducted a clever experiment by positioning the nerve of one nerve-muscle preparation on the intact muscle of another such preparation, stimulating the nerve of the second by means of an apparatus called an "inductonium." Both muscles contracted. Hence, the electronic measurement technique developed in close connection with knowledge in electrophysiology.

Emil Du Bois-Reymond (1818–1896), a German physiologist, postulated the existence of a potential difference between the interior and the exterior of a muscle and, consequently, of the membrane potential. In 1848, he was the first to detect electrical manifestations of voluntary muscular contraction in man ("Untersuchungen über Thierische Elektrizität," G. Riemer Verlag, Berlin, 1849). His method for detecting myoelectric activity was rather invasive, requiring removal of the skin with the goal of decreasing transfer resistance. The subject's fingers were immersed in a saline solution in which detection electrodes were also positioned and connected to a galvanometer. Du Bois-Reymond was probably the first to discover and describe that the production of contractile force in a skeletal muscle was connected with electrical signals originating in a muscle.[189]

An interesting illustration of the historical development of the "myograph" comes from Harry.[18] The instrument described registered the mechanical aspect of muscular function, i.e., contraction, directly. Because this chapter is concerned with methods of registering electrical phenomena paralleling the activity of skeletal muscles, an instrument of this kind would not, from the physical point of view, be appropriate. However, because Harry's survey very suggestively describes problems in the measurement of muscular function *in vivo*, it is worth mentioning due to the analogy with later developed methods of electromyography. Harry describes the first device of this kind according to Hermann von Helmholtz (1821–1894), a German physician, which was in fact created by accident since the motive was to determine latency time between the decision for voluntary muscular contraction and the actual initiation of mechanical muscular action. In the experiment mentioned, signal registration was provided by means of a kymograph according to Ludwig. (This is the graphic method applied by Marey, Section 1.2.) For the first time in the history of physiological measurements, a time resolution of 1 ms was achieved. Geddes and Baker[149] also used the instrument to measure muscular contraction based on a mechanical "angle-measurer" which uses an electrical potentiometer (similar to the electrogoniometer).

It is worth mentioning the "Group 1847," i.e., four researchers, Du Bois-Reymond, Ludwig, Helmholtz, and Ernst Brücke (1819–1892), who formed an intellectual coalition of sorts. They tried to change the popular belief in vitalism (the belief that living matter is in some way fundamentally different from non-living matter) into reductionism, i.e., the belief that the same physical laws apply to all of nature. This was of great importance to the development of modern science.

In the 1850s, the Frenchman G.B. Duchenne (de Boulogne) applied electrical stimulation to systematically research the dynamics and function of intact skeletal muscles ("Physiologie des mouvements," Librarie JB Baillieres, Paris, 1855, 1862, 1867). He described techniques and developed devices for registration and stimulation. In 1867 he was probably the first to pursue systematic research of muscular function using the electrical stimulation approach.

In 1868 Julius Bernstein first determined the waveform of neural action potential by using the rheotome (mechanical method, ballistic galvanometer). This was among the first applications of a sampling method by means of the Lenz commutator (Figure 6.1).[149] Augustus D. Waller (1887, according to Reference 188) intended to measure the latency period and the relationship between the action potential and force developed by m. gastrocnemius in the frog as a reaction to one stimulus. He

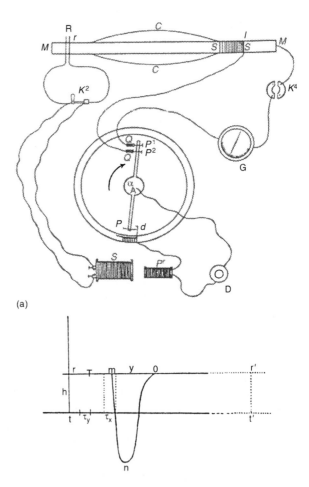

(a)

FIGURE 6.1 Bernstein's rheotome from 1876 for the measurement of the neural action potential which is one of the first applications of the sampling method. (From Geddes, L.A. and Baker, L.E. 1968. *Principles of Applied Biomedical Instrumentation.* New York: John Wiley & Sons. With permission.)

determined the existence of the latency, but its precise measurement was not possible until later by means of the following apparatuses: a micropipette electrode, a vacuum tube amplifier, and a cathode ray oscilloscope. The term electromyography stems from Marey ("Le vol des oiseaux," Mason G., Paris, 1890). In 1876 he developed a capillary electrometer. Dissatisfied with the inertia of this instrument, Willem Einthoven developed a string galvanometer (1903) based on Adler's telegraphic recorder for application in electrocardiography (1897) and won the Nobel Prize.

Basmajian and De Luca[51] consider the Englishman Arthur E. Baines to be the first biomedical engineer. Stressing the importance of technical considerations in connection with detection and interpretation of electrophysiological signals, Baines was the first to postulate the analogy between the spreading of the impulse through neural pathways and an electrical cable (1918). He modeled the nervous system on the theory of lines and circuits. (Such a line of thought was later developed in the work of Norbert Wiener, Lotfi A. Zadeh, and others in the development of systems theory and cybernetics.) Systematic and reliable measurements of myoelectric changes were, however, provided only by the construction of the metal surface electrode by the German Piper (1907).

Forbes et al. (1921, according to Reference 149) were perhaps the first to have applied the floating electrodes for measurements on a moving body: they measured EMG and ECG signals in elephants (!). They significantly improved methods of detection.

Forbes and Thacher (1920) used the cathode ray tube (invented by Braun in 1897) alongside a string galvanometer for amplification of action potentials. Requiring a device with a faster response, Herbert Spencer Gasser (1888–1963), an American physiologist, and Joseph Erlanger (1874–1965) used a Braunian tube, not a galvanometer, as a part of their circuit construction in 1922 and faithfully recorded neural action potentials. Their interpretation of the detected signals in the research of function of nervous fiber earned them the Nobel Prize for Medicine in 1944. Weddel and co-workers later introduced the method in clinical electromyography.[149] In 1937 the oscilloscope was developed.

The invention of the vacuum tube amplifier bears importance for the clinical application of electromyography and helped R. Proebster succeed in detecting the signal of a muscle diseased by peripheral nervous paralysis (Proebster 1928, according to Reference 51).[147] Furthermore, the concentric needle electrode invented by Adrian and Bronk (1929, according to Reference 51)[191] is very important for the clinical application of electromyography. Although the two researchers directed their work primarily at researching motor control and muscle schemes, this kind of electrode was of great importance for clinical application, since it enabled the detection of activity in individual (or small groups of) muscle fibers.

Adrian and Sherrington received the Nobel Prize for Physiology and Medicine in 1933 for discoveries in connection with the function of the neuron.

The work of Gasser, Erlanger, and Adrian significantly contributed to the understanding of the code of the nervous system, i.e., that intensity is expressed by the frequency of action potentials in one axon, while action potentials are mutually equal, and that intensity is also expressed by the number of axons which transmit information (pulse-frequency modulation, Section 2.2).

In Great Britain, Denny-Brown and Pennybacker (1939, according to Reference 187) distinguished between a nonvolatile (spontaneous) twitch of the inervated muscle and fibrillation of the denervated muscle by measuring electrical and mechanical muscular activity. They used subcutaneous needle electrodes, a mirror-type photographic recorder, and a Matthews oscillograph.

From 1942 to 1944, Herbert Jasper constructed the first electromyograph at McGill University (Montreal Neurological Institute). The RCAMC model (Royal Canadian Army Medical Corps) appeared in 1945. The first important electrokinesiologic study was conducted by Verne Inmann and co-workers (from the Berkeley Group) in 1944 and was directed at the shoulder area.[17] However, for kinesiological studies, the appearance of electrically stable electrodes (silver/silver chloride, Ag/AgCl) was of decisive importance. After World War II the field of myoelectric control of prostheses of the extremities began to develop in the former U.S.S.R. and the U.S., contributing significantly to development of the field of electromyography. EMG signals of the remaining musculature of a patient are used as control signals for the prosthesis, which is ideal, in principle, because natural neural control of motorics is imitated in this manner. An important contribution to the field came from Norbert Wiener at MIT, in connection with the concept of cybernetics. At a meeting of the Committee on Instrumentation and Technique of the American Association for Electromyography and Electrodiagnosis in 1954, the document "Information Concerning the Formulation of Minimal Requirements for Electromyographs for Clinical Use" was accepted. During the 1950s and 1960s, Buchtal and colleagues perfected the application of needle electrodes. Since the 1960s, electromyography has become a valuable clinical method in the diagnostics of muscular disease.

6.2   THE MYOELECTRIC SIGNAL MODEL

Strictly speaking, the myoelectric signal model should be presented after the conditions of its detection and measurement have been defined. However, since this book is concerned with measurement methods, the model will be presented prior to the description of the measurement technique itself, to attain a better understanding of the physiological process sensed by measurement. An article by Rozendal and Meijer[192] is relevant, in a sense, because it indicates certain circularity in logic when interpreting electromyographic findings.

Basics of neural conduction electrophysiology: The genesis of myoelectric phenomena was described briefly in Section 2.3 in connection with the biomechanical features of skeletal muscle. The process encompasses generation and propagation of action potentials in nerve and muscle cells related to the initiation and completion of (voluntary) muscular contraction. The corresponding electrophysiological phenomena are, in essence, based on the electrical conductivity properties of the corresponding membranes of the nerve, i.e., muscle cell, as well as on the characteristics of electrical disturbance which spreads inside the muscle cell. Equilibrium at the membrane, for a solution S, is generally described by the Nernst equation:

$$u_m = \frac{RT}{Z_i F} \ln \frac{[S]_1}{[S]_2} \tag{6.1}$$

where   R = gas constant = 8.3 J/mol K
        T = thermodynamic temperature in K
        $[S]_i$ = concentration of solution, i = 1, outside membrane, or 2, inside membrane
        Z_s = valency of solution
        F = Faraday constant $\approx 9.6 \cdot 10^4$ As

In a state of equilibrium, there is a constant electrical potential difference at the membrane $u_m = u_2 - u_1$ (u_2 denotes the potential inside and u_1 denotes the potential outside the membrane) with a value typically within an interval of −60 to −90 mV, i.e., the cell's interior is negatively polarized. The chemical mechanism by which such a value of potential difference is maintained is realized through the dynamic equilibrium of ionic currents (Na^+ and K^+ ions) passing through the membrane (the Na-K pump).[193]

Due to various kinds of stimuli, whether electrical, mechanical, or chemical in nature, the process of local depolarization may occur, i.e., a decrease in the negative potential difference at the cell membrane. Low-intensity stimuli cause a decrease in the polarization potential (voltage) after which equilibrium is reinstated. If the stimulus surpasses the membrane excitation threshold, a selective increase in membrane conductivity for Na^+ ions takes place (conductivity is a function of the membrane voltage) and, consequently, the flow of Na^+ ions toward the interior of the cell will increase significantly. The membrane conductivity for Na^+ ions in this working area is proportional to the degree of membrane depolarization, i.e., these two changes act in positive feedback. The process is continued until the cell's interior becomes electrically positive and the flow of Na^+ ions toward the cell's interior ceases. After this, the process of depolarization occurs and the membrane conductivity for K^+ ions increases, enabling an outward flow of K^+ ions. Then, there is a refractory period during which new repolarization cannot occur.

Figure 6.2 illustrates the theoretically reconstructed time curves of membrane conductivity (g) for Na^+ and K^+ ions and the corresponding changes of membrane electrical potential. This change of equilibrium potential at the membrane is called an action potential and is characterized by an impulse waveform (spike) of 0.2- to 1-ms duration and an amplitude of about 100 mV. If the surrounding membrane tissue is excitable and surrounded by a corresponding conductive solution, the action potential will induce its depolarization and this new change will begin to spread along the membrane as a regenerative chain process. The velocity at which the action potential spreads is a function of the membrane properties and its surroundings, i.e., the properties of the complete axon, and for neural action potential (muscular action potential will be considered later), values attained may be in the region of 0.5 to 120 m/s (Section 2.3).[47]

In general, the axon manifests its "cable" properties through two physical modes of signal transmission: passive and active conduction. The passive impulse conduction feature is realized when the active regenerative process of generation of the action potential is neglected. In this case, the membrane may be represented by a simple equivalent scheme as in Figure 6.3, where r_i [Ω/cm] denotes the axon's internal resistance per unit length and $1/r_m$ [Ω^{-1}/cm] denotes the membrane's conductivity per

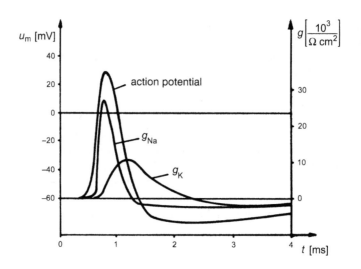

FIGURE 6.2   Theoretical waveforms of time dependencies of conductivity (g) of the cell membrane for Na$^+$ and K$^+$ ions and of the corresponding neural cell action potential. (From Katz, B. 1979. *Nerv, Muskel und Synapse*. Stuttgart: George Thieme Verlag. With permission.)

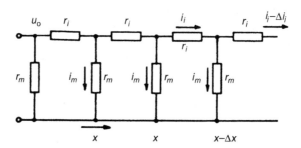

FIGURE 6.3   Equivalent electrical scheme of neural cell membrane for passive transport.

unit length. With voltage u_o at the beginning of the axon, the voltage at a certain distance x along the axon, approximately modeled as a cylinder, is determined by the expression:[55]

$$u = u_o \cdot e^{-x/\lambda} \tag{6.2}$$

where λ denotes the length constant with the value $(r_m/r_i)^{1/2}$. With this conduction mode, the depolarization wave is attenuated very quickly (after a few millimeters). If, however, conditions exist for realization of the aforementioned regenerative process of action potential generation (excitable membrane, conductive environment,

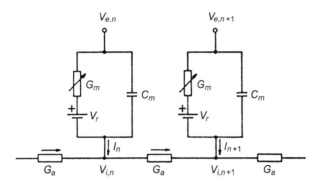

FIGURE 6.4 Equivalent electrical scheme of neural cell membrane for active transport. (From Vodovnik, L. 1985. *Nevrokibernetika.* Ljubljana, Slovenija: Fakulteta za Elektrotehniko. With permission.)

stimulus of a sufficient intensity), an equivalent scheme, such as in Figure 6.4, may be applied.

Neural fibers may be myelinated, i.e., may have the axon's membrane (myelin) acting as an electrical insulator. Action potentials are conducted along myelinated neural fibers by saltatory means, i.e., in "jumps." By regeneration at the Ranvier nodes, a high conduction speed is achieved with low energy expenditure. Transition from the nerve to the muscle action potential occurs at a motor point, i.e., plate (Figure 2.9).

Features of myoelectric processes: Muscle action potentials spread along a fiber from the stimulation point (motor point) in both directions, at a speed of 2 to 6 m/s, and are characterized by an impulse waveform, i.e., a spike with a duration of 1 to 5 ms (Section 2.3). A rapid passage of the muscle action potential is needed to secure a synchronous contraction of the whole fiber. Moreover, where the synchronism is especially important (fine motorics), the muscle is stimulated by multiple innervations by the same neuron, since neural action potentials travel faster than muscular ones. The described electrical processes for the spreading of the action potential in a neuron and then in a muscle fiber lead, with latency of a few milliseconds (latency time), to the realization of a muscle twitch. (Muscle twitch features, as well as the resulting macroscopic mechanical features of the muscle as a whole are described in Section 2.3.) Total muscle bioelectric activity is the result, corresponding to the nature of the physiological process of muscle force gradation, of the spatiotemporal activity of a large number of motor units (interference pattern). Myoelectric changes detected at the body's surface are additionally modified when passing through tissue and skin (filtering effect). This is a volume process spreading across a biological tissue according to the laws of propagation of electromagnetic fields in anisotropic media. The signal monitored at the body's surface is a result of the anatomic-physiological factors mentioned.

The waveform of a surface EMG signal as seen on the display of the registering device is further significantly influenced by the configuration of the detection electrodes and by the features of the amplifier. These technical factors, as well as some specific problems in connection with this (cross-talk), will be covered in the next section. It should once again be stressed that when considering the myoelectric signal model, the existence of detection electrodes is implicit to an understating of the model as they technically objectivize the signal.

Mathematical-physiological model of the myoelectric signal according to De Luca: Basmajian and De Luca[51] are of the opinion that the absence of a proper description of the EMG signal is probably the most important factor preventing electromyography from developing into a precise discipline. In their work, which summarized knowledge of and research results on muscular electrical function of the time and is an authority on this measurement method, these authors presented a mathematical model of the myoelectric signal. The model originally came from De Luca, who by systematizing numerous findings on the physiology of muscular activity as seen through myoelectric indices[194,195] comprehensively modeled generation and registration of the myoelectric signal and encompassed the complete chain from the motoneuron to the final amplified myoelectric signal at the output of the measurement device. The goal of this "structured" approach was interpreting information content of the EMG signal. Beginning with elementary physiological phenomena, the nerve and muscle action potential, all higher anatomical levels of integration were presented which finally result in the total EMG signal. The tissue filtering effect, as well as the influence of measurement instruments, were taken into account.

A neural action potential spreads along a motoneuron and, in a normal situation, activates all its branches; they in turn activate all the muscle fibers of a motor unit. If we assume, according to the usual practice of surface electromyography, bipolar detection and differential amplification of the myoelectric signal, the signal of each particular muscle fiber will manifest dependence of its phase and amplitude on the position of the fiber with respect to the detection electrodes. The amplitude of the action potential of a particular muscle fiber is a function of its diameter, its distance to the detection site, and the filtering properties of the electrodes. Ideally, in an infinitely long fiber and assuming there is an isotropic medium, Rosenflack (1969, according to Reference 51) has modeled the amplitude of an action potential by applying the expression $V = k \cdot a^{1.7}$, where a denotes the radius of the muscle fiber, while k is the constant. The value of k decreases in an approximately inverse proportion to the distance between the active fiber and the detection site. The duration of an action potential is inversely proportional to the speed of conduction in a muscle fiber (3 to 6 m/s). The filtering properties of bipolar electrodes are a function of the dimensions of the detection surfaces, the distance between the contacts and chemical properties of the metal-electrode interface (to be considered later). A situation in which there are more muscle fibers is illustrated in Figure 6.5, where the emergence of the motor unit action potential can be seen (the MUAP, motor unit action potential). A contracting motor unit changes its length, so a waveform of a MUAP changes accordingly. This anatomically complex situation is well described in Loeb and Gans.[196] Individual potentials may be monophasic, biphasic, etc. and their total sum represents motor unit action potential, denoted by h(t). (The term "phase" in electromyography

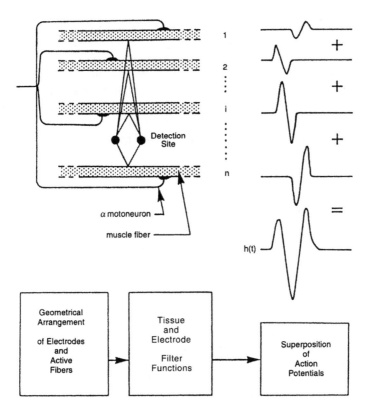

FIGURE 6.5 Schematic representation of the physiological model of motor unit action potential generation. (From Basmajian, J.V. and De Luca, C. 1985. *Muscles Alive—Their Functions Revealed by Electromyography.* Baltimore: Williams & Wilkins. With permission.)

denotes a region of positive or negative polarity of the action potential.[197]) Potentials may attain a triphasic form at best, while the appearance of more than four phases is indicative for pathology and a prolonged duration indicates the organism's aging.

Furthermore, a train of motor unit action potentials represents a MUAPT (motor unit action potential train). Considering the nature of the anatomical configuration of a muscle and the motor unit recruitment pattern in the course of contraction, only during contractions of minute intensity is just one motor unit activated, otherwise a number of them are active. MUAPT can be completely described by the waveform of the MUAP and interpulse intervals, IPIs. It can be represented as a random process where the firing rate is denoted by $\lambda(t,F)$. A systematic way of determining the mathematical expression for $\lambda(t,F)$ consists in fitting a histogram of interpulse intervals with the probability distribution function $p_x(x,f,T)$. Firing rate will be inverse to the mean value of $p_x(x_1 t_1 F)$:

$$\lambda(t,F) = \left[\int_{-\infty}^{\infty} x \, p_x(x,t,F) \, dx \right]^{-1} \tag{6.3}$$

MUAPT can suitably be broken down into a succession of Dirac delta impulses $\delta_i(t)$, passing through the filter (black box) whose impulse response is $h_i(t)$. A filter may be designed to have a time invariant response, to reflect any change in the waveform of the MUAP during a sustained contraction. De Luca refers to his papers in which such an approach has been elaborated, however, it is worth mentioning that this kind of modeling is restricted presumably by the existence of isometric contraction and is therefore an unrealistic model for locomotion situation.

A train of Dirac delta impulses may be described by the expression:

$$\delta_i(t) = \sum_{k=1}^{n} \delta(t - t_k) \tag{6.4}$$

Whereby MUAPT, $u_i(t)$, may be expressed as:

$$u_i(t) = \sum_{k=1}^{n} h_i(t - t_k) \tag{6.5}$$

where $t_k = \sum_{l=1}^{k} x_l$ for k,l = 1,2,3 . . . n

(n is the total number of IPIs in a MAUPT.)

MUAPTs, described by linear transfer functions $h_i(t)$, are summed up in space (determined by electrode configuration) and time, and such a signal, additionally filtered by passing through a biological conductor and contaminated by the noise of the electrode/amplifier system, finally results in the total EMG signal.

The model of the EMG signal is synthesized by linearly summing up the MUAPTs:

$$m(t,F) \sum_{i=1}^{p} u_i(t,F) \tag{6.6}$$

which is schematically presented in Figure 6.6.

However, when the electrode generates a large number of MUAPTs (larger than 15), typical for surface electromyography, then the law of large numbers may be revoked and a simpler and more limited approach may be applied. It is then efficacious to model the EMG signal as a limited frequency band signal with Gaussian amplitude distribution. Also, the amplitude may be represented by a carrier signal whose features are those of the signal during isometric contraction, and a modulating signal reflecting force- and time-dependant signal features. Basmajian and De Luca refer to Kreifeld and Yao (1974) and Hogan and Mann (1980).

A representation complementary to the one in the time domain is via the signal power density spectrum, estimated by means of the Fourier transform:

$$S_m(\omega) = \sum_{i=1}^{p} S_{ui}(\omega) + \sum_{\substack{i,j=1 \\ i \neq j}}^{q} S_{uiuj}(\omega) \tag{6.7}$$

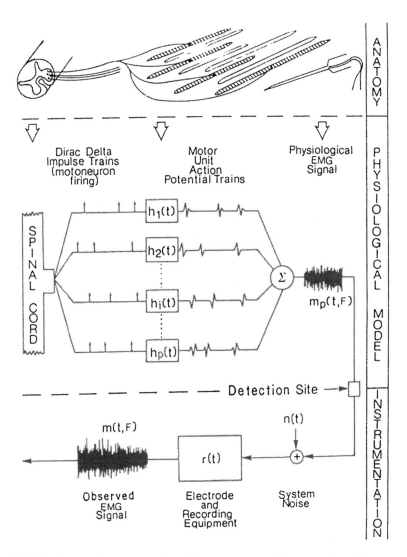

FIGURE 6.6 Schematic representation of the generation and formation of a global myoelectric signal. (From Basmajian, J. V. and De luca, C. 1985. *Muscles Alive—Their Functions Revealed by Electromyography.* Baltimore: Williams & Wilkins. With permission.)

where $S_{ui}(\omega)$ denotes power density spectrum of the i-th motor unit's action potential train $u_i(t)$, while $S_{uiuj}(\omega)$, a cross-spectrum of trains $u_i(t)$ and $u_j(t)$, denotes contribution to the total spectrum density of those motor units whose firing frequencies correlate to a certain degree (q, the number of those motor units). Here, it holds that $q \leq p$. During isometric contraction, firing frequencies of active motor units correlate to a significant degree (De Luca et al. 1982, according to Reference 195).

Taking into account the filtering influence of the electrode/amplifier system on a myoelectric signal, the power density spectrum of the final signal registered may be expressed as:

$$S_m(\omega, t, F) = R(\omega, d) \left[\sum_{i=1}^{p(F)} S_{ui}(\omega, t) + \sum_{\substack{i,j=1 \\ i \neq j}}^{q(F)} S_{uiuj}(\omega, t) \right] \qquad (6.8)$$

where d denotes the interelectrode distance.

Basmajian and De Luca[51] have analyzed the subject of EMG signal spectral representation in conditions of isometric contractions. Relying also on the spectrum model by Lindström (1970), which presumes a cylindrical form of the muscle fiber, they discuss the interdependence of impulse propagation speed along a fiber, the distance of detecting electrodes, and consequent changes to the frequency spectrum. This opened up a broad field of study which has experienced considerable development recently, especially in Sweden and the U.S. During conditions of isometric muscle contractions, the characteristics of spectral changes with the occurence of local muscle fatigue are reliably determined: frequency components shift toward lower values and signal component amplitudes increase. The median and mean frequency have been established as the most significant spectral parameters by which the phenomenon of fatigue may be characterized and observed. (As these issues are related almost exclusively to conditions of isometric muscle contractions, they will not be discussed in further detail.) The study of changes in the spectral representation of the EMG signal during dynamic contractions is only beginning to be explored systematically, but not in natural locomotion. Difficulties in satisfying the assumption of the quasi-stationarity of the signal are a significant problem, so in principle, a suitable approach for dynamic muscular activities, typical to locomotion, is using time-frequency analysis. Potentially suitable are mathematical models such as the one determined by the Wigner-Ville distribution or based on wavelets, etc. One approach, in which a simple cyclic flexion-extension movement of the lower leg is analyzed during exercise on a training device by healthy subjects (athletes), has been researched and an original laboratory method has been developed for measurement and time-frequency spectrogram analysis of surface EMG signals for the upper leg musculature to determine fatigue.[197]

It is obvious that the statistical properties of the time chains of motor unit action potentials are decisive for the resulting features of the total signal. As has been said, a large number of technical factors contribute to the signal's appearance which also depends on detection and amplification. This will be explored in the following sections. The amplitude of resulting surface EMG signals has stochastic properties and can be well described by the Gaussian distribution function. Amplitude ranges from 0 to 6 mV peak-to-peak or 0 to 1.5 mV RMS.[198,199] The frequency spectrum width is 10 to 500 Hz, while the dominant components are in the interval of 50 to 150 Hz. In human locomotor measurement signals, the EMG signal poses the strongest requirements on the A/D conversion procedure.

Myoelectric changes are related to muscle force, but this connection is not simple, nor is it of a linear character. It is influenced by various physiological factors, such as conditions (eccentric, concentric, isometric) and speed of contraction, instantaneous muscle length, the state of local muscle fatigue, specificities in muscle and body structure of the respective subject, specificities in the muscle measured with respect to others, etc. Efforts to quantify this relationship, i.e., to determine the correlation between the EMG signal and muscle force, will be discussed in Section 6.3.3.

6.3 KINESIOLOGICAL ELECTROMYOGRAPHY

6.3.1 SURFACE ELECTROMYOGRAPHY

Electromyography is the detection, amplification, and registration of bioelectrical activity changes in the skeletal musculature. This method may be applied on the surface (metal disk electrodes) and under the surface, either subcutaneously (wire electrodes) or intramuscularly (needle or wire electrodes).[200] In clinical neurology and physical medicine, different variants of electromyographic measurement techniques are routinely applied in the diagnostics of particular neuromuscular pathologies. Only surface electromyography, i.e., the detection and measurement of muscular action potential changes manifested at the surface of the skin, above the measured muscle, will be covered. This subgroup of electromyographic measurement techniques is the one most often applied in locomotion measurements.

This method is suitable for the study of motion, primarily since it is noninvasive. By positioning detection electrodes on the skin above the surface of the monitored muscle, global insight into the resultant bioelectric activity is gained, as opposed to the application of spatially significantly more selective intramuscular electrodes. Also, in surface electromyography, a larger degree of measurement repetitiveness in successive trials (smaller run-to-run variability) is achieved (Kadaba et al. 1985, according to Reference 7) than is the case in the application of needle electrodes. Basmajian[201] (in 1968, the opinion at that time), however, lists a limitation of surface electromyography as being mostly suitable for studies of large skeletal musculature, located near the body's surface, as opposed to the application of intramuscular electrodes which, according to this author, enable precise scientific applications of the method.

Measuring EMG signals during movement is called dynamic electromyography by some authors, this being the only method in practice capable of determining which muscles are active and when during a certain movement. Applications of the EMG measurement technique in studies of human movement and locomotion have stimulated development of the field known as electrophysiological kinesiology. In 1965, ISEK (the International Society for Electrophysiological Kinesiology) was founded and corresponding measurement, protocol, and methodological standards were established.[202] The society holds meetings on a biennial basis. A specialized periodical has been issued by ISEK, the *Journal of Electromyography and Kinesiology,* since 1991. It should be stressed that the EMG measurement technique is by far a subjective skill, an inherent ability or "trade," evident from the actual narrative style of some referent books (Loeb and Gans,[196] for instance). Besides this work, and that mentioned by Basmajian and De Luca,[51] in the description of this measurement method several references were relyed upon.[146,147,149,191,198,199,204,205]

6.3.1.1 Surface Electrodes in Electromyography

Electrodes are the sensor elements in the measurement chain, making contact with the subject's body to detect bioelectric changes; in this way they act as a kind of electrochemical transducer for reactions occurring within the living organism. Because of this contact, by definition, the requirement for the measurement method to be non-invasive is disturbed to a certain degree. However, as will be seen, the complete procedure, beginning with the contact between electrode and skin, should not significantly disturb the subject's perfomance or measurement and contact itself does not really set limitations in this case, but possibly other factors do (e.g., cables).

Electrical conductivity in the human body is accomplished by means of ions as charge carriers. The detection of bioelectrical changes, therefore, means interaction with ion charge carriers and transformation of ions into electrical currents, which is accomplished through electrodes, electrical conductors in contact with the ionic solution of the body acting as a receiving antenna.[196] Regardless of the type of electrode, the physical-chemical basis for the sensor procedure is the properties of the metal-solution interface with the solution in contact with metal. The current is carried by free electrons in the metal and by ions in the solution, so a corresponding transformation occurs, whereby the interaction between electrons in the electrode and ions in the solution significantly influences the performance of such sensors. At the interface of an electrode and an ionic solution, redox (oxidation-reduction) reactions occur, necessary for transporting the charge between the electrode and the solution. In general, these reactions are described by equations:[203]

$$C \rightleftharpoons C^{n+} + ne^-$$
$$A^{m-} \rightleftharpoons A + me^-$$

(6.9)

where n denotes the valence of cation material C and m denotes the valence of anion material A. In most electrode systems, cations in a solution and the electrode metal are the same, and by releasing electrons, C atoms oxidize and become part of the solution as positively charged ions. These ions are reduced when the process occurs in the reverse direction. In the case of anion reaction (the second expression), the directions for oxidation and reduction are reversed. For the electrode to function optimally, these two reactions must be reversible, i.e., both directions should be equally easy to realize.

Interaction between metal in contact with the solution and metal ions in the solution results in a local change in the concentration of ions in the solution in the region close to the surface of the metal. This causes disturbance in the neutrality of charges in this region. As a result, the electrolyte surrounding metal gains a different electrical potential from the rest of the solution. A potential difference is thus established, called the half-cell potential or electrode or polarization potential. This potential, in effect, is a result of the difference in the velocity of diffusion into and out of the metal.[146] By creating a charge layer at the interface, a balance is formed. (In the case of a semipermeable membrane surrounded by solutions, as in a cell membrane, the Nernst potential mentioned is established according to Expression 6.1.) This charge, in fact, is a double layer (electrical double layer) where the layer closer to the metal has one polarity and that closer to the solution has another, with a pronounced capitive action, 30 μF/cm^2.[205] The value of the half-cell potential, consisting of a DC and an AC component, depends on temperature, sweating, and relative movements between the metal and the skin, as well as the electrode input current. Various materials are characterized by different values for the half-cell potential: for instance, for Ag + Cl$^-$ it is + 0.223 V and for Au it is + 1.680 V. Various attempts have been made to stabilize the value of the half-cell potential through construction of the electrode.

A perfectly polarizable electrode would function so that a current passes between the electrode and the electrolytic solution, thereby changing charge distribution in the solution in the vicinity of the electrode. Therefore, the actual current, then, does not pass through the interface electrode-electrolyte. In practice, this situation is approximated by noble metals. A perfectly polarizable electrode, on the contrary, would allow the current to pass freely through the electrode-electrolyte interface without changing the charge distribution in the electrolytic solution in the vicinity of the electrode. In practice, the best material for electrodes with these approximate features has proven to be silver/silver chloride (Ag/AgCl).

Direct contact between the metal and the solution changes during movement, so the features of the region described also change, i.e., the electrochemical voltage is variable, so one speaks about the movement artifact. In practice, therefore, usually the detection surface of the electrode makes contact with the skin through electrolytic gel (conductive paste), which decreases transfer resistance and also acts as an adaptable mechanical contact.

The electrode design described by Piper (1912) has not significantly changed to this day, and metal material is still used. In EMG measurements, standard silver plate electroencephalographic (EEG) electrodes may be used, however, the most suitable are Ag/AgCl electrodes (chlorided electrodes according to Šantić[205]), in which an appreciable decrease of the AC component of the half-cell potential is achieved. Furthermore, it is worth mentioning that in an effort to minimize electrode mass, a layer of silver-metal film on the skin is used (applicable in astronauts) as a detection surface, but this kind of solution does not allow signal registration of good quality. Flexible electrodes made out of either a fine net of silver wire or rubber which becomes conductive by adding carbon may be used.[205] The specific resistance of these electrodes is 15 Ωcm, making them suitable for kinesiological applications. Based on long-term research and experiments in electromyography, De Luca suggested the following electrode design as being optimal: the electrode is in the form of a strip and not circular. His argument was that electrodes in the form of strips with an area equivalent to a circular electrode crosses the direction of a larger number of muscle fibers, resulting in a stronger signal (superior S/N) which is more representative for more fibers (factor 2.8). However, further consideration will presume electrodes circular in shape for the sake of convenience. The impedance of surface electrodes is usually between 2 and 10 kΩ.

An Ag/AgCl electrode consists of a silver base structure coated in a layer of ionic compound of AgCl. Because AgCl is relatively unsoluble in a water solution, this surface remains stable. Also, the half-cell voltage of this electrode is minimal, so the movement artifact is smaller. The influence of frequency on electrode impedance is also smaller, especially at low frequencies, and electrical noise is also smaller than in a polarizable electrode. Electrodes of this type are manufactured by Beckman in the U.S. They are of a very low mass, 250 mg and 11 mm in diameter, highly reliable, and durable.

The DC component of the half-cell potential is electrically neutralized by using electrodes in pairs, in a bipolar arrangement. In this case, there are two half-cell potentials, so the total voltage equals their difference. Figure 6.7 depicts the equivalent schematics for the bipolar arrangement. In practice, perfect symmetry is impossible to achieve.

In preparing the measurement procedure, prior to filling the electrode with gel and positioning, the chosen site of contact must be cleaned with alcohol and even slightly abraded to decrease skin resistance as much as possible. As a rule, interelectrode resistance should be kept below 10 kΩ. (In practice. the amplifier input impedance should be at least ten times the electrode impedance, in order to transfer a small bioelectrically generated current accurately. There are special electrolytic gels incorporating cleaning ingredients, and they can be applied prior to electrode positioning without special skin treatment.) First, an adhesive collar is positioned at a chosen site with the hole located at the intended location of contact. Then, the electrode must be

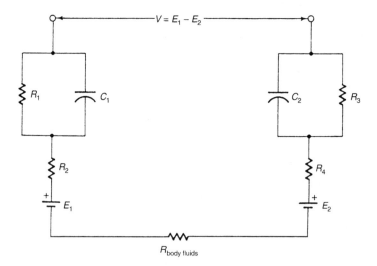

FIGURE 6.7 Equivalent electrical scheme of the bipolar electrode connection. (From Cromwell, L., Weibel, F. J., and Pfeiffer, E. A. 1980. *Biomedical Instrumentation and Measurements.* Englewood Cliffs, NJ: Prentice-Hall. With permission.)

positioned. After this, additionally it must be mechanically attached using adhesive tape. Figure 6.8 shows the positioning of surface EMG electrodes manufactured by Beckman to dominant lower leg musculature with the goal of providing kinesiological measurements. Two signal-electrodes, with a 30-mm interelectrode distance, are positioned standardly, approximately at the midrange of the muscle, along muscular fibers, i.e., parallel to the voltage gradient being measured. In addition, a neutral electrode is also needed, if possible with a larger contact area which serves as mass. It is sufficient for all measuring channels and is positioned at a distant, neutral site on the body with respect to voltage (in this case, the ankle). After the measurement session, electrodes have to be cleaned properly. (Beckman provides the complete procedure.) The procedure chosen is in accordance with the ISEK recommendation[202] and has been used in measurements by the author.

There is an on-going debate among experts, however, regarding the actual positioning of the electrodes for kinesiological measurements with regard to the muscle. The conservative opinion regarding two signal electrodes was that they have to be positioned at the midpoint, the most prominent part of muscle (e.g., Nilsson et al.,[150] at a distance of 15 to 30 mm). A more exact approach to electrode positioning, however, presupposes that the location of the motor point (plate) has been determined beforehand. This is accomplished by electrically stimulating the muscle and determining the location of stimulation where the muscle has the greatest mechanical response. For a long time, the opinion held was that electrodes should be positioned as close as possible to the motor point.[5,196,206] Loeb and Gans[196] explained this traditional attitude: their opinion is that "electrodes have to be positioned reasonably close

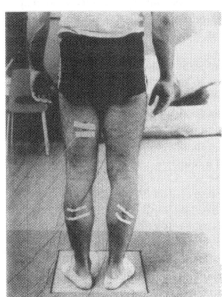

FIGURE 6.8   Positioning of surface EMG electrodes on major lower extremity muscles for kinesiological measurements.

to the motor point with the goal of obtaining a signal of maximum and constant amplitude." However, from the point of view of signal stability, this location is the worst. In this region, the action potentials travel caudally and rostrally along muscular tissue consisting of fibers, so the positive and negative phases of the action potential are mutually neutralized. Basmajian and De Luca, therefore, are of the opinion that electrodes must be located approximately at the midpoint between the determined motor point location and the point where the muscle and tendon join because signal properties are the most stable there.

As far as interelectrode distance is concerned, De Luca and Knaflitz[199] recommended a value of 10 mm, center to center: the interelectrode distance influences signal spectrum width and signal amplitude; decreasing the distance results in a shift of the spectrum toward higher values and signal amplitude decreases. It is therefore necessary to keep the distance fixed, to allow quantitative comparisons of measured values intra- and intermuscularly, as well as between subjects. (Detection surfaces should, if possible, be attached to the fixed surface.) A 10-mm distance is considered to be a good technical compromise because a representative electrical muscle activity is detected during contraction (several cubic centimeters of muscular tissue), while the unwanted signal (common mode, cross-talk) is limited. However, in measuring dynamic muscle activity, it is often impossible to keep the interelectrode distance constant, which introduces additional variability to the measurement procedure. It is customary to locate the third, neutral electrode as far away from the muscle as possible. As an addition to experiment documentation, and in order to achieve repeatability of the measurement procedure, it is a common practice to take a photograph of the actual electrode placement. However, the dilemmas mentioned remain. The *Journal of Electromyography and Kinesiology* offered reccomendations for reporting EMG measurements (Appendix 2) in which, however, electrode positioning is not prejudiced.

In the Beckman Dynograph measurement setting, the electrodes so positioned have to be connected via 460-cm (15-ft) cables to the preamplifier inputs, which are part of the instrument. In multichannel measurements, typically on 6 EMG channels, the cable system should not disturb the subject and it allows registration of multichannel measurement records of a succession of performances. Our experience supports this possibility of measuring, not only simple, but also relatively complex movements, provided they are performed over a sufficiently small area, such as backward somersaults in gymnastics or tennis serves. Certainly, these measurements require collaborative and patient subjects because contact may be lost easily.

Henneberg[191] emphasizes, correctly, the basic principal limitation confronting a person measuring EMG signals. Although this primarily refers to needle detection, it is useful to repeat Henneberg's point of view. A large quantity of information on the muscle's bioelectric condition is contained in the time variable spatial distribution of electrical potentials in a muscle tissue. Unfortunately, it is not practical to detect, by adequate resolution, 3D spatial samples of potential distribution, since it would require sticking hundreds of electrodes into a muscle. To decrease discomfort as much as possible for the subject, routinely only one electrode inserted into the muscle is used. In this way, detection of a signal is realized in which the signal represents a time change of potential in not only one point of the muscle tissue, but along the muscle fiber as well. In analogy, the surface electode detects the time curve of potential change which represents the global sum of the one real volume process. Henneberg stresses the importance of the skill of the person measuring this process since a 3D process is deduced from a 1D quantity "on the basis of intuition."

6.3.1.2 Myoelectric Signal Amplifiers

Multiple sources of noise are present when measuring EMG signals. The most important are

1. Surrounding sources of electromagnetic radiation, of which the most prominent is power line voltage of 50-Hz (or 60-Hz) frequency and its harmonics which are electrostatically or magnetically coupled to the body (This is the dominant source of noise with an amplitude of one to three orders of magnitude larger than the EMG signal.)
2. Unwanted biopotentials generated in other bodily sources
3. Movement artifacts which are a consequence of the electrode-skin interface on the one hand and moving cables on the other (These sources cover a region from 0 to about 20 Hz.)
4. A broadband noise characteristic of electronic components for signal amplification and processing

The whole procedure, therefore, must be such that successful detection and amplification of myoelectric signals are achieved in the environment described. EMG signals are of small amplitude, while source impedance is high, as previously stated. Detection electrodes are connected to the preamplifier input. To amplify myoelectric signals, differential amplifiers are used, suitable for signal amplification in the presence of noise, since they amplify the difference between two signals, detected at two detection sites (bipolar detection). In this way, useful information, a signal relatively close to the detection electrodes, is selectively amplified, while noise, made by the total signal from sources relatively distant from the detection site (all previously mentioned influences), which is approximately equal at both detection sites, is canceled. In this way, differential amplifiers realize good separation of signal from noise. The accuracy with which the procedure is carried out is expressed by the common mode rejection ratio (CMRR). The rejection factor of an amplifier is defined as a ratio between the amplification factors of the differential mode and the common mode. Figure 6.9 depicts the block scheme of the differential amplifier. The following expression holds:

$$U_{out} = G_D U_{biol} + \frac{G_D V_C}{CMRR} + G_D U_C \left(1 - \frac{Z_{in}}{Z_{in} + Z_1 - Z_2} \right) \qquad (6.10)$$

G_D is the amplification factor of the differential mode.

The output of the real amplifier consists of a wanted output component and an unwanted component produced by imperfect rejection of the interference signal (CM

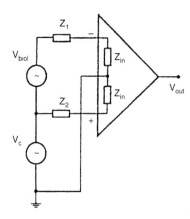

FIGURE 6.9   Block scheme of a differential amplifier. (From Nagel, J. H. 1995. *The Biomedical Engineering Handbook,* J. D. Bronzino, Ed. Boca Raton, FL: CRC Press, 1185–1195. With permission.)

interference signal) as a function of the rejection factor and the unwanted component is the result of nonequality in source impedances (asymmetry), which allows a small part of the CM signal to appear as a differential signal in the amplifier. Ideally, $Z_1 = Z_2$, and the rejection factor's value is ∞. The values of a rejection factor of 32,000 (90 dB) usually suffice for unwanted noise to be rejected. Modern biopotential amplifier solutions have a rejection factor from 120 to 140 dB.[207] Measurements without special protection systems are therefore possible (unlike situations in monopolar measurements of high precision, where a Faraday cage is needed). Gain should have a value in the range of 100 to 50,000, with signal/noise ratio as high as possible. Gain should be such that the output signal amplitude of an order of V is achieved. Noise must be $<5\ \mu V$ RMS. Input impedance must be $>10^{12}\ \Omega$ in parallel to the 5-pF capacitance. The frequency spectrum should be from 20 to 500 Hz, according to the spectral characteristics of surface EMG signals.

   Figure 6.10 is the scheme of all major aspects of EMG signal acquisition, while Figure 6.11 illustrates a summary of corresponding filter effects present.

   To protect the individual from electrical shock, isolation is provided via either optical means or an isolation amplifier, and corresponding construction standards for electromedical equipment have to be fulfilled.

   A classical solution for the instrument is a rack-mounted, robust construction, used traditionally in electromyography (see Figure 6.17). This is the Beckman Dynograph type of recorder: an 8-channel measurement and registration device produced in the U.S. since the 1970s and laid out in discrete electronic technology. The rejection factor accomplished is ≥ 100 dB at 60 Hz and ∞ at DC, the input resistance is $\geq 2\ M\Omega$, the equivalent input noise with a short circuited input is 1 μV RMS, sensitivity is 1 to 5 μV/mm, and frequency response is adjustable to 10 kHz, 150 Hz (used for averaged EMG), 30 Hz, and 0.3 Hz. The upper limit frequency of the pen (curvilinear coordinate system) is 150 Hz. Modular construction of the device and a large number of different coupler units for various other bioelectric, kinetic, and

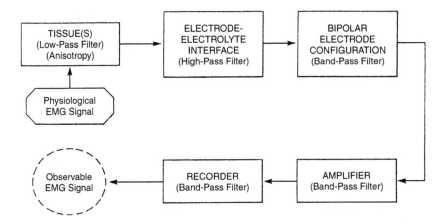

FIGURE 6.10 Principal aspects of surface EMG signal acquisition. (From Basmajian, J.V. and De Luca, C. 1985. *Muscles Alive—Their Functions Revealed by Electromyography.* Baltimore: Williams & Wilkins. With permission.)

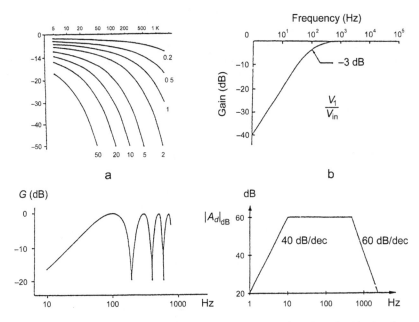

FIGURE 6.11 Filtering effects of tissue (a), metal-electrolyte interface (b), bipolar electrode configuration (c), and electronic instrumentation (d), which all influence the EMG signal. (From De Luca, C.J. and Knaflitz, M. 1992. *Surface Electromyography: What's New?* Boston: Neuromuscular Research Center. With permission.)

biological signals make the device very suitable for application in biomechanical and neurophysiological labs. Measurements of ground reaction force by the strain gage-based force platform (Chapter 5) also included this device with appropriate coupler units. The device's outputs are connected to a PC by means of an A/D converter.

An ideal differential amplifier is not found in practice, so there is always some asymmetry. The electrode cables are coupled capacitively to the power supply. In this situation, both signals of differential and of common modes are generated. In an effort to eliminate these kinds of errors, active electrodes were developed, whereby the preamplifier is within the electrode's immediate vicinity (Figure 6.12). Unity gain amplifiers of small output impedance are employed for this purpose. In this way, they act as impedance transformers, bringing features of the electrodes closer to those of an ideal voltage source. Active electrodes are therefore characterized by much higher input impedance than passive ones, by virtue of which they are significantly less sensitive to the (variable) impedance of the electrode-skin layer. As a result, preparation of the skin and the use of conductive gel is eliminated (dry electrodes). Coupling to the skin of such electrodes may be resistive or capacitive in character. In the latter, the detection surface is covered by a thin layer of dielectrical material. The capacitively coupled electrodes, however, suffer from a higher inherent noise level and dielectric of variable properties. De Luca designed and constructed an active electrode with a high input impedance amplifier, 10^{12} Ω, and a small capacitance, 3 to 4 pF, in a junction field effect (JFET) microelectronic technology. Two dry electrode constructions in monolithic technology with a built-in amplifier were described by Šantić.[205] Besides giving an EMG signal of superior fidelity, active electrodes are very practical for diverse applications; however, their disadvantage is that they introduce noise to the measurement chain; since noise appears at the very beginning of a chain, its influence is the greatest. However, in the majority of applications, imported noise is negligible.

Advancements in electronic technology have made possible construction of miniaturized and compact ergonomically designed EMG devices of a Holter type.

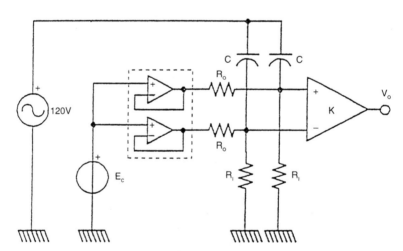

FIGURE 6.12 The active electrode principle. (From De Luca, C.J. and Knaflitz, M. 1992. *Surface Electromyography: What's New?* Boston: Neuromuscular Research Center. With permission.)

For instance, the ME3000 Professional (manufactured by the Finnish firm Mega Electronics Ltd.) weighs only 600 g including batteries, is $166 \times 77 \times 30$ mm in size, and allows measurement and storage of up to 4 surface EMG channels, with a sampling frequency of 2 kHz per channel and long recording duration. Its small mass and size make the device very practical, so that it can be used in measuring various movement structures, as long as they are not too fast, such as sprint-like locomotion. For instance, EMG activity during the long jump in track and field sporting events cannot be measured successfully because during the approach run, the device fixed to the athlete's body trembles forcefully, thereby certainly hindering the athlete's concentration and thus becoming invasive.

The ME3000 Professional uses self-adhesive electrodes (which include a gel-supplied wad), Blue Sensor type SE 00 S, by the Danish manufacturer Medicotest. Cables (1.56 m in length) which connect the electrodes to the device have built-in instrumentation preamplifiers (gain 375) and filters, and their housings also act as neutral electrodes. They are connected to the third electrode next to each measurement channel; therefore movements of the measurement electrodes are minimized as are movement artifacts. The input impedance is 10 GΩ. Rejection factor is $\geq$ 130 dB, input noise is $<$ 1 μV in the frequency band 20 Hz to 480 Hz, S/N is 60 dB, and input dynamics is $\pm$ 6 mV. The lower limit frequency is 15 Hz (a first-order high-pass filter), and the upper limit frequency is 480 Hz (a fifth-order low-pass filter). A built-in A/D converter has 12-bit resolution, and the sampling frequency is adjustable (per channel) to the following values: 250 Hz, 500 Hz, 1 kHz, and 2 kHz.

The device can function in a direct (on-line) mode and is connected to the PC by a 2.5-m optical cable (RS-232 serial input), so that measurement signals are directly transferred into the PC in real time. It can also function in a Holter mode (data logger), and signals are stored on the memory card and transferred to a computer later (off-line). Internal memory has a 1 MB capacity and memory cards (PCMCIA) have 1 MB as well.

In multichannel measurements, the problem of cross-talk always occurs. Therefore, it is necessary to take into account the actual functional-anatomical characteristics of the skeletal musculature monitored during the measurement procedure to minimize cross-talk.[208]

6.3.2   EMG TELEMETRY

In analogy to the trend encountered in human kinematics measurements (the problem of body markers), it is also desirable for EMG measurements to be as noninvasive as possible. Within this context, and to enable subjects to move freely both indoors and out, there is a need to eliminate wires, i.e., cables, in electromyography. Complete lack of contact is, however, impossible to achieve because actual detection electrodes are needed, as are the remaining miniaturized elements of a measurement device. Among bioelectric signals generated by the human organism (EMG, ECG, EEG), it is precisely the EMG technique for which the possibility of telemetry is the most important because it is most frequently applied to the moving human body.

Measuring of free and natural movements has been considered the true prerequisite for research into the neuro-musculo-skeletal system and complex human motorics.

There are various technical possibilities to realize multichannel EMG measurements and subsequent signal radio transmission to the stationary device for display and data storage. Some of them will be described.

Rositano and Westbrook (1978, according to Reference 147) describe a broadband EMG telemetry system. The spectrum width is 20 to 2000 Hz. Because EMG requires such a spectrum width and more channels (12 here), it would be difficult to implement a multiplex by using one radio frequency (RF) link in a device with a weak power supply and reasonable reach. A broadband device was therefore developed consisting of miniature, crystal-controlled RF transmitters, positioned at the site of each electrode. Working frequencies are attributed to transmitters in the range 174 to 216 MHz. They are small enough to be located at the sensor sites and have a linear frequency response of 20 to 2000 Hz and a 15-m reach. Receivers are crystal-controlled, enabling precise channel identification. Amplifier input impedance is 10 MΩ. There is no need therefore for cable connection from the electrodes to the common transmitter, so greater mobility is achieved.

Welkowitz et al. described the 15-channel biotelemetry system used for EEG signals (originally from Deutsch (1976), according to Reference 209), using frequency modulation and time multiplex. This is a one-RF carrier system, frequency modulated by means of time multiplex signal. The maximum sampling rate for one channel is 400 Hz, so it can only be used for average EMG.

Noraxon is a Finnish firm (1986), now active in the U.S. (Scottsdale, Arizona). The EMG Myosystem offers both classical and EMG telemetry with excellent properties, ranging to 1000 m. Telemyo allows measurements of 1 to 32 channels of a raw (spectrum width not specified) or integrated EMG signal, with adjustable channel sensitivity, a UHF band width of 400 to 470 MHz, with transmission in binary form, 250 Kbit/s, with 12-bit amplitude resolution. Including the battery, the mass of the transmitter unit is 859 g. The IECV-601 standard is satisfied. EMG telemetric devices are appreciably more expensive than conventional "wired" devices.

Appendix 2 contains standards for reporting EMG data from the *Journal of Electromyography and Kinesiology*.

6.4   MYOELECTRIC SIGNAL PROCESSING

The surface EMG signal is quasi-stochastic, the amplitude ranges approximately from 0 to 6 mV, and it has a frequency spectrum between 10 to 500 Hz. EMG signals may be analyzed in their raw form, but mostly only qualitatively. In order to represent measurement information in the most appropriate way and to ease its interpretation, various EMG signal processing methods have been developed. In the recommendations by ISEK,[196] basic processing methods in domains of time and frequency are presented. Developments in the field of electronic and computing technology offer a very good basis for the efficacious practical realization of various signal processing methods. Problems which occur relate to signal modeling (Section 6.2). The trend toward changes can be seen by comparing two editions of Basmajian's reference book within a time span of 18 years (1967 and 1985): the newer edition contains a completely new chapter on mathematical signal modeling. This is especially true today in view of original processing methods being developed on the basis of, for example, neural networks, applicable at the level of modeling the neuromuscular system.[210]

6.4.1   TIME DOMAIN PROCESSING METHODS

Figure 6.13 illustrates standard methods for processing the EMG signal in a time domain (ISEK).

Full-wave rectification: In this procedure, the entire energy of the signal is retained and the mean value is other than zero, allowing the application of various averaging procedures. (Because the EMG signal is typically recorded through an AC coupled amplifier, direct application of the averaging operation would result in a zero value. This also holds true for DC coupling if the DC polarization potential equals zero.)

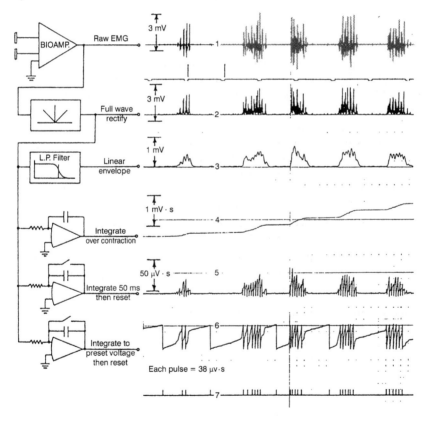

FIGURE 6.13   Common surface EMG signal processing methods in the time domain. (From Winter, D.A., 1980. *Biomechanics and Motor Control of Human Movement, 2nd Edition.* New York: John Wiley & Sons. Reprinted by permission of John Wiley & Sons, Inc.)

Averaging (smoothing) of the full-wave rectified signal: The procedure may be realized by analog or by digital means and consists of suppressing higher frequency components, i.e., low-pass filtering of a full-wave rectified EMG signal. The filter spectrum width determines the degree of smoothing. Equivalent to smoothing, in the digital sense, is averaging of the full-wave rectified signal (averages, means rectified signal).

$$\overline{|m(t)|}_{tj - ti} = \frac{1}{t_j - t_i} \int_{ti}^{tj} |m(t)| \, dt \tag{6.11}$$

$T = t_j - t_i$ represents a time window. When the window is moved along the whole time record, the operation is called "moving average." There are various possibilities for positioning the window. Typically, the T value is between 100 and 200 ms.

The most common type of EMG signal processing in a time domain is full-wave rectification followed by some form of smoothing, i.e., low-pass filtering, also called averaging. According to Hof (1980, according to Reference 211), such an approach is compatible with the following EMG signal description:

$$e(t) = I(t) \cdot n(t) \tag{6.12}$$

with n(t) denoting a stationary stochastic process with a zero arithmetic mean and a unity variance. I(t) denotes the time-variable intensity of EMG and e(t) is the recorded EMG signal. Such an approach is in correlation with the signal model, but is a crude simplification.

In practice, averaging is by far the method most often used. In older instruments, such as the Beckman Dynograph, it is accomplished in an analog form, using a linear RC filter (time constant, 100 ms). In more recent instruments, it is by digital means (Mega, Noraxon). In spite of the simplicity and practicality of the analog method, it is clear that, in principle, a superior processing procedure is when the raw signal has been digitized previously and then smoothed by digital means since, if smoothing is provided twice consecutively in both directions of the time axis, this enables the elimination of phase lag (as introduced by analog filtering). The smoothed EMG signal is primarily used because, in a way, it represents a certain correlate of muscle force (although the influence of elasticity of muscles and tendons "is not seen," only active generated force).

Integration: Integration means the calculation of the area below a curve, and it is also applied to the full-wave rectified signal. The operation therefore results in a value expressed in Vs (mVs). In professional reference books, the term has often been used incorrectly; it is, in fact, a "linear envelope detector."

$$I\{|m(t)|\} = \int_0^t |m(t)| \, dt \tag{6.13}$$

It concerns a subgroup of operations for obtaining "average rectified value"—except there is no T.

Root mean squared (RMS) value:

$$\text{RMS}\{m(t)\} = \left(\frac{1}{T}\int_t^{t+T} m^2(t)\,dt\right)^{1/2} \tag{6.14}$$

Calculation of RMS value is a type of processing recommended by Basmajian and De Luca, the reason being that RMS represents a signal power for EMG signals detected during voluntary contraction and has, as opposed to the results of the remaining processings, physiological meaning.

Number of zero crossings: This time-domain operation consists of counting the crossings of the EMG curve through the zero line. In this way, the calculation of the medium of the signal spectrum is approximated, which should otherwise be calculated by means of FFT (the next section). This type of procedure was once of great value because the FFT algorithm was not easily accessible or realizable.

6.4.2 PROCESSING IN THE FREQUENCY DOMAIN

Spectral representation of the EMG signal is estimated by digital means by FFT algorithms. The dominant field of application of such a representation is in the evaluation of local muscle fatigue due to isometric contraction, not so much in studies of locomotor activities. The most significant spectrum parameters for the evaluation of muscle fatigue are considered to be its median f_{med} and mean value f_{mean}:

$$\int_0^{\omega_{med}} S_m(\omega)\,d\omega = \int_{\omega_{med}}^{\infty} S_m(\omega)\,d\omega \tag{6.15}$$

$$\omega_{mean} = \frac{\int_0^{\infty} \omega\, S_m(\omega)\,d\omega}{\int_0^{\omega} S_m(\omega)\,d\omega}$$

6.4.3 NORMALIZATION

Absolute EMG signal amplitude values (expressed in mV) are influenced by factors such as skin filtering influence, etc., so in repeated measurements on the same subject (electrode repositioning), it is not possible to realize reliable comparisons (intertrial or intermuscular). Furthermore, comparisons of values of a certain muscle in different subjects are also not possible on an absolute scale. Therefore, in kinesiological measurements, it is customary to normalize the signal amplitude in some way. The amplitude of the signal measured during maximal voluntary isometric contraction of the corresponding muscle is chosen (MVC) as the value to which the normalization is made (100%). However, what the biomechanical conditions should be when determining this value (i.e., the value of the angle in a corresponding joint) is a question which remains to be answered. However, this is not the only amplitude normalization method because individual researchers use their own modifications.[211] Normalized EMG signals enable valid intersubject and intermuscular comparisons and analyses. (Of course, in certain experimental situations, direct comparisons of signals expressed in absolute units, volts, are also possible.) When cyclical locomotor activities are being measured, typically gait, an EMG record may also be normalized along the time coordinate to the duration of one cycle (period) (also valid for the remaining locomotion measurement variables).

6.4.4 EMG SIGNAL AS AN ESTIMATE OF MUSCLE FORCE

Since the first papers were published in the 1950s, many researchers have published papers dealing with the subject of the relationship of EMG signals to muscle force. Various types of muscular contractions, various muscles, etc. have been researched and there are important survey papers.[49,212-215]

Bouisset[49] (the Laboratorie de Physiologie du Mouvement, Universite de Paris-Sud, Orsay, France) reasons that since the EMG signal represents a measure of the muscle's excitation, and the force depends directly on the excitation, the EMG represents an effective indirect measure of force. He analyzes the relationship force/EMG in conditions with no fatigue and concludes that consistent relations exist between the surface integrated (Expression 6.13) EMG signal and various quantities which characterize the mechanical functioning of muscle. This also holds true for both isometric and anisometric contraction.

Perry and Bakey[212] supplement Bouisset's survey based on experience at the Pathokinesiological Laboratory of the Rancho Los Amigos Hospital in Downey, California, where gait has been analyzed since the 1960s with emphasis on kinesiological electromyography. (This is a continuation of the California School of Inmann and Sutherland.) Considerations refer to voluntary movements and a healthy neuromuscular system. While these authors have established the existence of a certain relationship between the EMG and strain in the domain of isometric contractions, in the domain of arbitrary movement, such as occurs in locomotion, there is no simple relation. In specific tasks, such as isotonic or carefully controlled unidirectional movements, there is evidence of the existence of relationships that may be mathematically expressed in a simple form. However, generally, due to the dynamics of moving limbs and nonlinearities inherent in force/velocity curves (Section 2.3), simple expressions do not exist in anisometric cases. (The situation is different in an artificially stimulated muscle which will not be considered here.) Skeptical toward the possibility of good mathematical models existing which would describe the relationship EMG/force, and based on an extensive literature review, Perry and Bakey conclude that:

- There is no universally accepted method of measurement, processing, and quantification of EMG signals.
- During isometric contractions, a linear proportionality exists among a corresponding quantified EMG measure and the registered force, at least in the vicinity of the midrange of operating force, and for some electrode configurations.
- The proportionality constant between the quantified EMG signal and force depends on joint position (i.e., on muscle length).

- During movement, the relationship EMG/force cannot be described by linear algebraic equations.
- The choice and location of electrodes and the applied processing methods have a significant role in the evaluation of the EMG/force relationship.

6.4.5 MULTICHANNEL EMG SIGNAL PROCESSING

Apart from their illustrative "descriptive" role, multichannel EMG signals may serve as a kind of "window" into the function of the neuromuscular system, since they represent correlates of its functioning. Multichannel EMG signal processing may be carried out on raw signals, but it is more suitable to do this on previously processed EMG signals, namely, in the first approximation the average EMG may be considered to be a muscle force correlate. Such waveforms are amenable to processing by correlation analysis.

Gandy et al.[216] have proposed a method of processing averaged EMG signals by which the degree of mutual connection of bioelectric activity of the two muscles measured is estimated. The degree of connection of bioelectric activity is estimated by calculating the coefficient of correlation between the two signals according to the expression:

$$\text{c.c.} = \lim_{T \to \infty} \frac{1}{T} \int_{-T/2}^{T/2} x(t)\, y(t)\, dt \tag{6.16}$$

where x(t) denotes average EMG signal of the first and y(t) denotes average EMG signal of the second muscle. The range of possible resulting values is between 1 and -1, where 1 denotes an extreme case of maximum positive correlation, 0 denotes the lack of correlation, and -1 denotes a case of maximum negative (inverse) correlation (e.g., in antagonist muscles). Such a method has been applied by the author and his collaborators and it will be presented in the next chapter.

Shiavi and colleagues worked on the methodology of processing multichannel averaged EMG locomotor signals.[217–219] Methods such as Karhunen-Loewe expansion were applied, with the goal of reducing the dimensionality of the feature vector, and particular methods of feature extraction and clustering were also applied in order to classify EMG patterns.

6.5 APPLICATIONS OF SURFACE ELECTROMYOGRAPHY

Examples of the application of surface electromyography in human locomotion measurements are presented.

6.5.1 WALKING AND RUNNING

Mochon and McMahon (1980 and 1981, according Reference 22) suggested the ballistic walking model of human gait based on the inertial features of lower extremities' motion and with minimal muscular activity which, in a way, may be considered a continuation of the historical approach by the Weber brothers. Generally in walking, there is very weak activity in the upper and lower leg muscles compared to volitional movements registered.[51] According to this, the musculature would primarily act passively due to its own mass and inertia and not actively; however, this may be considered as only the first approximation of the real situation.[73]

Motivated primarily by requirements of rehabilitation medicine, walking, a fundamental locomotor activity, was electromyographically researched most. The typical lower extremity myoelectric activity patterns in normal gait are established. Figures 6.14 and 6.15 illustrate examples of multichannel records of raw and processed EMG signals in gait. The support and swing phases of the extremity are present in a complete stride. During the support phase, beginning with the first heel strike, maximum activity is first attained by the hip joint extensor muscles: biceps

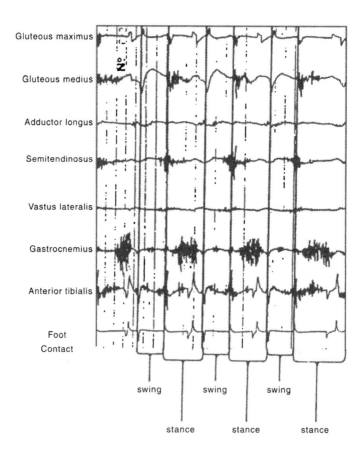

FIGURE 6.14   Lower extremity EMG activity in normal walking. (From Basmajian, J.V. and De Luca, C. 1985. *Muscles Alive—Their Functions Revealed by Electromyography.* Baltimore: Williams & Wilkins. With permission.)

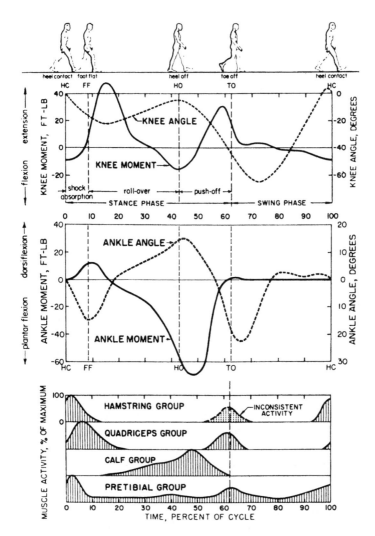

FIGURE 6.15   Processed EMG signals and some kinematic and kinetic quantities of lower extremities in walking. (From Basmajian, J.V. and De Luca, C. 1985. *Muscles Alive—Their Functions Revealed by Electromyography.* Baltimore: Williams & Wilkins. With permission.)

femoris, semitendinosus and semimembranosus (the hamstrings), and the ankle flexor tibialis anterior. Immediately after this, as body weight is transferred over the supporting leg, knee extensor muscles rectus femoris and vastus intermedius attain maximum activity, to stabilize the knee joint. At the end of the support phase, during foot lift-off, the ankle extensor group is activated: m. gastrocnemius and m. soleus. During toe lift-off, (weaker) upper leg muscle activity reoccurs, while the hip extensor group is not activated consistently. The tibialis anterior is also active; this muscle shows weak bioelectric activity during the entire stride. During the swing phase, the remaining muscles are silent. Analogously, muscles of the other leg also act with time shift. The complete walking cycle consists of two- and one-leg support phases.

Muscular activity during the support phase is mostly isometric, while mandatory activity during the swing is mostly isotonic, but in reality, as already mentioned, muscular contractions in natural locomotion are always of a combined isometric-isotonic nature.

Many authors (University of California at Berkeley, 1953, according to Reference 220) generally claim that EMG signals manifest good reproducibility in multiple trials by the same subject at a certain walking speed. However, there are variations among subjects. Amplitude dispersion of the averaged EMG signal may be seen in m. gastrocnemius (Figure 6.16), derived from measurements on six subjects. Changes in EMG activity of this muscle in different walking modes can also be seen. Moreover, in healthy individuals walking at a spontaneous speed, a few distinct multichannel EMG activity patterns have been established in terms of significantly different time synchronisms.[208] This indicates a certain redundancy in the design of the neuromuscular system, representing a basis for various possible dynamic realizations of approximately the same movement structure.

When walking speed is increased, myoelectric activity rises as well[221] in all muscles except the m. tibialis anterior. This muscle manifests a local minimum in EMG activity, somewhat more pronounced in some individuals, at speeds of 1 to 1.5 m/s. This could, therefore, be "the most comfortable" (natural, spontaneous) walking speed during which energy expenditure is minimal. At speeds above 1.8 m/s, bioelectric activity rises suddenly in all muscles as does energy expenditure.

At "natural" walking speeds, the antagonist muscles do not act simultaneously.[222] This indicates an effort toward a more rational expenditure of energy, i.e., its minimization during this kind of locomotion. In most of the muscles (semitendinosus, semimembranosus, biceps femoris, rectus femoris, iliacus, gastrocnemius, and soleus), the beginning of electrical activity coincides with maximum length[222] (in concordance with Gelfand and Tselin 1971, Kostyuk 1974, Gurfinkel et al. 1973, all according to Reference 222). Under these conditions, muscle stimulation by the motor neural paths is increased due to the action of muscle spindles sensitive to stretch (facilitation or potentiation reflex via I_a afferent neurons, see Figure 2.7).

If walking speed is further increased to about 2 m/s, walking becomes running. A double support phase no longer exists, and there are periods when the entire body is in the air, without floor contact. Transition is accompanied by restructuring of the muscular activity pattern, reflected in the EMG signals. A larger differentiation in walking vs. running is found in the ankle joint: during running there is a time overlap in the activity of the foot flexors and extensors, m. tibialis anterior and m. gastrocnemius. The purpose of such a simultaneous contraction of agonist muscles is for improved stabilization of the lower extremities at an increased movement speed. (Nilsson et al.,[150] treadmill running). When running speed is increased, an anticipation of activity in m. biceps femoris, m. semitendinosus, and m. semimembranosus, prior to heel strike, is observed.[222]

Basmajian and De Luca[51] described the bioelectric activity of the complete leg and trunk musculature during normal locomotion, dominantly gait, but also running and walking up and down stairs, etc. Their approach is descriptive and typically functional-anatomical. Nilsson[25] synthesized published results of EMG measurements in gait and running. (It should be pointed out that there is a significant tradition of

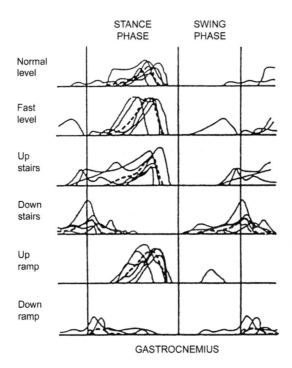

STANCE   SWING
PHASE   PHASE

Normal
level

Fast
level

Up
stairs

Down
stairs

Up
ramp

Down
ramp

GASTROCNEMIUS

FIGURE 6.16 Processed EMG signals of m. gastrocnemius in different modes of walking for six examinees. (From Berme, N. 1980. *Biomechanics of Motion,* A. Morecki, Ed. Berlin: Springer Verlag, 185–217. With permission.)

electromyography measurements in Sweden, especially in medical rehabilitation and ergonomics, but also in basic research of motorics and neurophysiology in humans and animals. Interpretations are mostly descriptive.) Regarding the clinical applicability of the described measurements, Perry[223] made concise and suggestive arguments for the use of EMG measurements in various gait pathologies and gave examples of measurement signals and their clinical interpretation, which offers a solution to the problem of proper differential diagnostics and follow-up and the evaluation of therapy. Her interpretations of EMG signals are also descriptive, with simple processing for determining the state of muscle (on-off).

Forssberg and colleagues[54] moved away from the descriptive level in myoelectric definitions of gait. Studying the evolution of this basic motor stereotype, they determined the term of man's "plantigrade" gait (Section 2.2), a mode of locomotion differentiating man significantly from other animal species. Using a combined measurement method including kinematics, some kinetic quantities, and EMG signals, they described the following determinants of "plantigrade" gait by relying on traditional kinematic gait determinants (Saunders et al.,[87] Section 4.2.1):

• Heel strike phase
• Knee flexion during the support phase

- Asyncronous EMG pattern
- Rotation, tilt, and translation of the pelvis

An important feature of "plantigrade" gait is control of foot-ankle movement. A powerful heel strike on the ground is developed primarily due to active ankle dorsiflexion during knee extension at the end of the swing phase. This is accomplished by a prolonged activity of the m. tibialis anterior during swing. Heel strike induces an ankle extensor moment, causing foot plantar flexion as opposed to the eccentric contraction of m. tibalis anterior. The contraction of the quadricep(s) muscle in heel strike results in knee extension during foot contact, and its prolonged activity during the support phase enables smooth knee flexion, damping vertical body oscillations. Forward body transfer is produced by a forceful contraction of the calf musculature, extending the ankle joint in the second half of the support phase. A time sequence of joint movements, ankle extension followed by knee flexion and hip flexion, reflects active forward propulsion due to the action of ankle extensors. A specific action in each joint creates an asynchronous pattern of interjoint coordination in which joint movements occur in mutual phase shift and where flexion and extension are initiated at different time instants. Forssberg and colleagues described the kinematics of the pelvis and trunk in detail. Patterns of muscular activity, reaction force, moments, and movements in joints are the result of neural activity in the CNS and its effect on the musculoskeletal system. Biomechanics in man differs considerably from the biomechanics of other mammals (including primates) and this may explain some of the differences found. But, pronounced differences in the pattern for the activation of leg musculature suggest a considerable reorganization of the central circuits during the phylogeny of man's "plantigrade" gait.

Surface electromyography as a technique therefore enables detailed insight into muscular activity during gait and thereby corresponding insight into the function of the neuromuscular system. (However, one has to be aware of limitations due to the small number of measured channels and because of accessibility to surface muscles only.) The EMG signal as a "window" into the action of the CNS is an illustrative analogy. By recognizing patterns of multichannel EMG signals, it may be possible to identify basic neurophysiological mechanisms in the peripheral neuromuscular system—reciprocal inhibition, for instance, is a typical example (Section 2.2, Basmajian and De Luca[51]). Shiavi and colleagues have developed mathematical-statistical procedures for the analysis of multichannel EMG signals in order to describe and classify measurement samples, intended for application in clinical work. Efforts to automatically extract the phases of muscular bioelectric activity from EMG signals recorded during gait was described in Shiavi et. al.[219] It concerned an EMG activity envelope feature extraction scheme. The procedure recognizes time, duration, and amplitude of phases of the activity. The method is based on a Tauberian approximation in which curve waveforms are modeled as the sum of identically shaped pulses with different time lags and amplitudes. The authors selected an average linear envelope as the representative EMG signal sample. The amplitude and duration of stride normalizations are provided. Data from a data base for six muscles were used (tibialis anterior, gastrocnemius, rectus femoris, medial hamstring, gluteus medius, and lateral hamstring) in 20 adults at 4 gait velocities. The raw EMG signal was rectified by analog means and integrated by means of the Paynter integrator of the third order with an equivalent time integration constant of 12.5 ms. The averaged curves obtained, called linear envelopes, are sampled with a 250-Hz frequency, and time base interpolation transforms the time scale of each step into 256 points. Interpolated linear envelopes of a step are averaged and pattern average is obtained. Amplitude normalization is provided according to Yang and Winter (1984, according to Reference 219), by means of a within-subject ensemble average. The technique is sensitive enough to be able to discriminate between two phases occurring quickly in succession.

Based on the development of EMG signal processing methodology, Chen and Shiavi[217] developed an automatic clustering technique for linear EMG envelopes during gait which uses representation of temporal features and a maximum peak matching scheme. Adjustment of the envelope is provided by dynamic programming, giving qualitatively the largest numbers of adjusted peak values and quantitative measurement of the minimum distance. The resulting averaged EMG profiles have a small statistical variation and can serve as a template for EMG comparison and further classification. In this way, a discrimination is possible between normal and pathologic subjects. The system was tested on 30 healthy individuals and 15 individuals with cerebral paralysis, ages ranging from 4 to 11 years. Tibialis anterior, peroneus longus, gastrocnemius, soleus and tibialis posterior muscles were measured. It was determined that, even in normal individuals, a number of patterns exist for a given muscle. There is no sharp limit between normal and pathological for a particular muscle, and so to solve the problem an analysis of the synergy action of the neuromuscular system through multichannel EMG is needed.

Lower extremities are only the most prominent body parts during locomotion, especially gait and running. Many investigations have been conducted by measuring trunk surface myoelectric activity in these locomotions. Except for pointing to two relevant references,[51,224] the results of this research will not be described.

6.5.2   BACKWARD SOMERSAULT IN GYMNASTICS

Electromyographic measurements have been performed to research take-off abilities in gymnasts (Section 5.1.4.2). The activity of major lower extremity muscles (m. gastrocnemius, m. tibialis anterior, m. rectus femoris, m. biceps femoris) was monitored during performance of a backward somersault from a standing position. This gymnastic technique element was chosen for two reasons: first, it enabled insight into the level of acquired performance skill because it concerns a complex movement structure; and second, it was possible to measure technically using the available, classical wire surface EMG method. Gymnasts require a number of years of training to acquire the backward somersault skill. This element represents a significant component of the performance repertoire. The study was confined to the sagittal plane, according to the schematic representation of the musculoskeletal system of lower extremities in Figure 2.4. A method for measurement, processing, and interpretation of EMG and take-off force signals was developed, aimed at objectivizing the diagnostics of this fast, skilled movement stereotype.[73,160,225,226] The method encompassed these subunits:

- Determination of the movement structure, in this case a backward somersault from a standing position
- Determination of muscle equivalents[49] for the chosen movement structure and its representation in the activity space (The activity states of particular muscles are eccentric, concentric, isometric contraction, and inactivity.)
- Realization of measurement with simultaneous grading of performances (Figure 6.17 presents a female subject, a young gymnast in top condition, with electrodes positioned, prior to the performance of a backward somersault from a standing position.)
- Identification of time phases of the movement structure on measurement records of particular performances
- Calculation of bioelectric, kinetic, and derived indexes expressing the degree to which the motor skill has been acquired

Equipment which has been described previously, hardware- and software-wise (Chapters 3, 5, and 6), was used. Software support in BASIC was upgraded for processing measured kinetic and EMG signals. It allowed realization of a group of

FIGURE 6.17 Female subject with electrodes positioned and fixed, ready to perform the backward somersault from a standing position, take-off from the platform (Biomechanics Laboratory at the Faculty of Physical Education in Zagreb).

processings of multichannel registered signals and the calculation of correlations with other criteria used. Scores by gymnastic coaches for each performance trial were used as a criterion for the quality of the performance of the backward somersault from a standing position. Surface EMG signals were detected, amplified, and averaged on-line by analog means, i.e., full-wave rectified and low-pass filtered (analog RC filter, 100-ms time constant), which is a part of the Beckman Dynograph device ("averaged" measurement mode). The upper frequency thus attained was 150 Hz. Signals were further digitized on-line and stored in computer memory. Quantification according to Gandy et al.[216] was applied (Expression 6.16).[73,160,225]

Experiments yielded the following quantitative criteria for the level of skill acquisition in the performance of a backward somersault from a standing position. The kinetic criterion for good-quality performance is determined by values of the impulse width of the vertical force $F_z < 300$ ms and of ratio $F_{zmax}/BW > 3$, while the bioelectric criterion is determined by the value of the correlation coefficient of averaged EMG signals of the left and right m. gastrocnemius measured to the amount of ≥ 0.8, reflecting a high level of symmetry in the activity of ankle extensor muscles. (BW denotes body weight.) Figure 6.18 presents raw registered EMG signals and ground reaction force in a sagittal plane, and Figure 6.19 presents the corresponding averaged signals generated by a high-quality, top-level gymnast. (Recordings do not belong to identical, but only to similar performances.) Figure 6.20 illustrates a measurement record of the performance of a physical education student who mastered the movement structure 3 weeks prior to measurement as part of the regular gymnastics curriculum. Visually, differences may be observed from the recording of the previous figure: muscle activity is bilaterally symmetrical to a lesser degree and, in addition, the waveform of take-off force is less impulsive. Analysis of results of a larger subject sample confirmed these observations. Furthermore, Figure 6.21 illustrates the record of the same top-level gymnast in a deliberately poor performance. Subtle differences may be observed. This example indicates the high sensitivity of the method in the determination of fine changes in motorics manifested, i.e., in the level of the acquired movement skill.

The bioelectrical criterion for this method of trained movement quantification has been further elaborated into the moving correlation function:[227]

$$H(j) = \frac{\sum\limits_{i=1}^{200} [A(i+j) - A][B(i+j)]}{\left\{ \sum\limits_{i=1}^{200} [A(i+j) - A]^2 \, \Sigma[B(i+j) - B]^2 \right\}^{1/2}} \qquad (6.17)$$

The function H(j), a collection of scaled correlation coefficients, calculated one by one for each j, shows the correlation between two selected averaged EMG signals. It is calculated by moving a 200-point window A over the original 300-point function B starting from the " −50 point" to the " + 50 point."[228,229] The function H(j) thus has 100 points in total (j = 100) with an expected maximum around or at the "50 point." Figures 6.22 and 6.23 illustrate calculated moving correlation functions for the "top-level" and "poor" performer and for performances by a top-level performer "at his

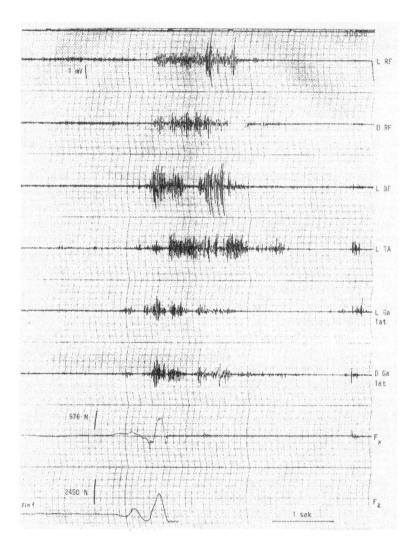

FIGURE 6.18　Multichannel EMG recording of the backward somersault performed from an initial standing position. Major lower extremity muscles are monitored. (Note the curvilinear plot.) Abbreviations as follows: L RF, left m. rectus femoris; R RF, right m. rectus femoris; L BF, left m. biceps femoris; L TA, left m. tibialis anterior; L Ga, left m. gastrocnemius; R Ga, right m. gastrocnemius; F_x, horizontal component of ground reaction force in sagittal plane; F_z, vertical component of ground reaction force. Male subject, active in sports gymnastics.

best" and "deliberately poor," respectively. A good discrimination feature is observed in the procedure for the evaluation of skill level realized in this way. The method is also potentially applicable in the quantification of acquisition of other movement structures (this presumes that other muscles are measured) and might also serve in monitoring the progress in motorics for particular problems in rehabilitation medicine.

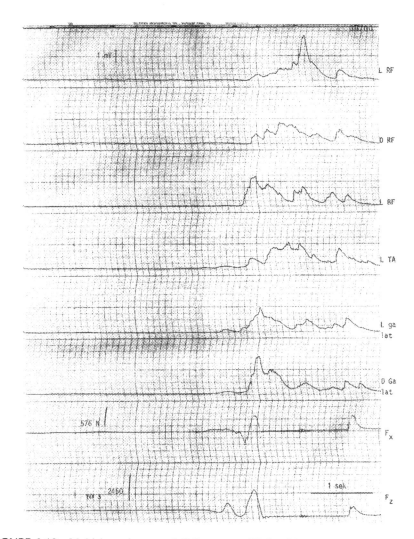

FIGURE 6.19 Multichannel averaged (full-wave rectified and low pass filtered) EMG signals and takeoff force in the sagittal plane during performance of the backward somersault from a standing position. (Note the curvilinear plot.) Male subject, a top-level sports gymnast. Same symbols as in Figure 6.18. (From Medved, V. and Tonković, S. 1991. *Med. Biol. Eng. Comput.* 29:406–412. With permission.)

To conclude, the following are qualities of the surface electromyography method in the study of human locomotion:

- The advantage of unique information content (i.e., this is the method which enables noninvasive insight into the dynamics and energetics of muscular activity, limited to registering active contractile component of muscular force)

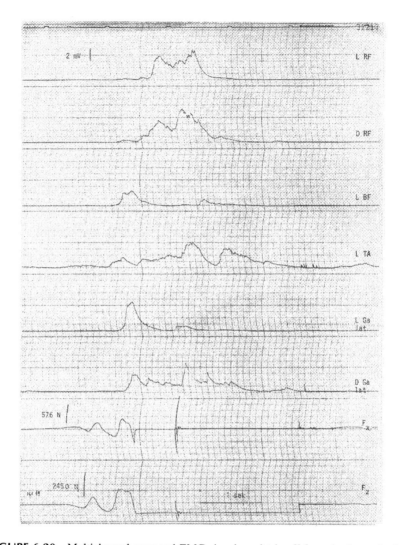

FIGURE 6.20 Multichannel averaged EMG signals and takeoff force in the sagittal plane during the performance of the backward somersault from a standing position. (Note the curvilinear plot.) Male subject, a student of physical education. Same symbols as in Figure 6.18. (From Medved, V. and Tonković, S. 1991. *Med. Biol. Eng. Comput.* 29:406–412. With permission.)

- The disadvantages of
  - Inaccessibility of subsurface-located musculature
  - Relative invasiveness, evident, to a degree, in electrodes, but above all in cables, limiting the application area significantly
  - Telemetric devices which are rather expensive and more difficult to access for average working environments (A compromise is the Holter EMG method.)

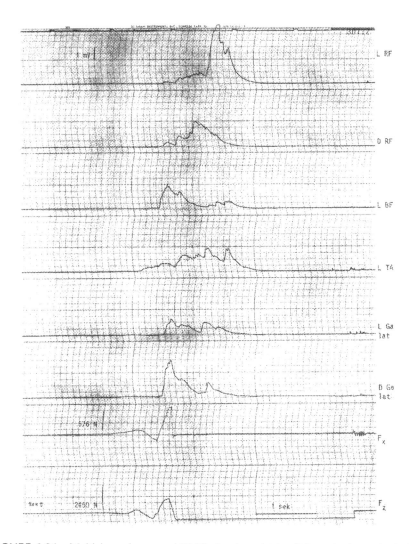

FIGURE 6.21   Multichannel averaged EMG signals and takeoff force in the sagittal plane during performance of the backward somersault from a standing position. (Note the curvilinear plot.) Male participant, a top-level sports gymnast when performing "deliberately poorly." Same symbols as in Figure 6.18. (From Medved, V. and Tonković, S. 1991. *Med. Biol. Eng. Comput.* 29:406–412. With permission.)

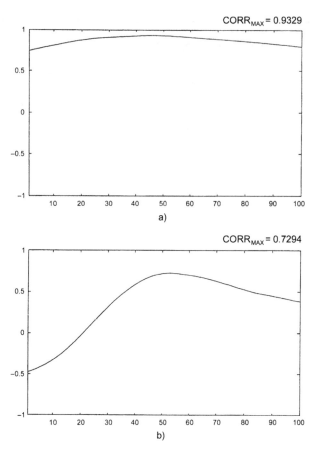

a)

b)

FIGURE 6.22 Correlation function. An electrophysiological criterion of acquisition of a complex sports movement structure. Top-level vs. "poor" performer of the backward somersault based on averaged EMG signals of left and right m. gastrocnemius. (From Medved, V., Tonković, S., and Cifrek, M. 1995. *Med. Prog. Technol.* 21:77–84. With kind permission from Kluwer Academic Publishers.)

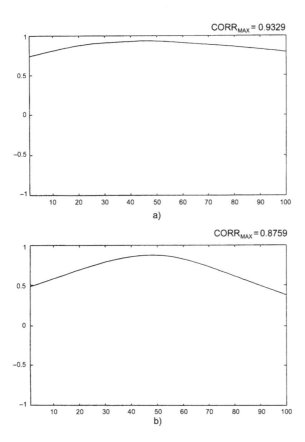

FIGURE 6.23   Correlation function. "At his best" vs. "deliberately poor" performance of a top-level performer of the backward somersault based on averaged EMG signals of left and right m. gastrocnemius. (From Medved, V., Tonković, S., and Cifrek, M. *Med. Prog. Technol.* 21:77–84. With kind permission from Kluwer Academic Publishers.)

7 Comprehensive Locomotion Diagnostic Systems and Future Prospects

In principle, only by simultaneously measuring kinematic, kinetic, and EMG quantities can an experimental paradigm for the study of locomotion be accomplished (see Chapter 2). Some of the results in previous chapters reflected correlation of diverse measurement quantities: EMG signals and ground reaction force or EMG signals and kinematic quantities, for instance.

This chapter presents certain essential practical features of some contemporary commercial solutions to measuring a number of locomotion variables, with corresponding processing and presentation of data (Section 7.1). The goal is not to provide a comprehensive survey, but to offer an appendix of sorts of the material previously presented, picking a few complex solutions for the measurement and/or diagnosis of locomotion. This completes the methodological approach in Chapter 2. The focus is on solutions accessible to the user and their clinical application and use in biomechanical diagnostics of movement—a field which is becoming a standard in medical and sports/training environments. Also a description of the initial development phases of an automated, completely noninvasive kinematic measurement method is given; this approach does not require use of body markers of any kind (Section 7.2).

7.1. COMPREHENSIVE BIOMECHANICAL MEASUREMENTS AND CLINICAL APPLICATIONS

The world market of locomotion measurement systems offers solutions aimed at completeness and comprehensiveness. A summary of the practical features of some of them will be given.

Oxford Metrics Ltd. (U.K.) offers a comprehensive solution of the VICON measurement system. At the beginning of the 1990s, the system was used in more than 100 labs in more than 20 countries in gait analysis in medical rehabilitation, orthopedics, sports, ergonomics, and in industry. The VICON System 370 encompasses all three groups of locomotion measurement data and is supported by user software (VICON Clinical Manager Software). At least four cameras are needed for unilateral measurement and five are needed for bilateral measurement; it can accommodate a maximum of seven cameras and is IBM PC compatible. Up to four platforms can be incorporated. The MA-100 EMG system, by the same manufacturer, has ten EMG channels and an additional eight channels for foot switches. As part of installing the system, the manufacturer offers to mount the platform in the floor of a laboratory.

When using the system, the entire process, from system calibration to user-tailored reports, is controlled via a PCX monitor, in a Microsoft® Windows® environment. According to the manufacturer, conditions for the photogrammetric procedure and biomechanical analyses are rigorously adhered to (see Kadaba, M.P., Ramakrishnan, H.K, and Wootten, M.E. 1990. Measurement of lower extremity kinematics during level walking. *J. Orthop. Res.* 8, 3:383–392 and Davis, R.B., Õunpuu, S., Tyburski, D.J., and Gage, J.R. 1991. A gait analysis data collection and reduction technique. *Hum. Movement Sci.* 10(5): 575–587).

The system allows kinematic and kinetic measurements, an inverse dynamic approach, gait analysis, stabilometry, calculation of energy transfer between segments, measurements and analysis of EMG signals. Figure 7.1 illustrates a gait cycles window by which data on ground reaction force, EMG signals, and a 3D stick diagram, representing a reconstructed kinematic video record of body movement superimposed on the platform and with ground reaction force vector representation, can be rotated in space to realize a view from any one angle, and this can be scrolled forward and backward. Figure 7.2 is a typical monitor display of a gait report. This

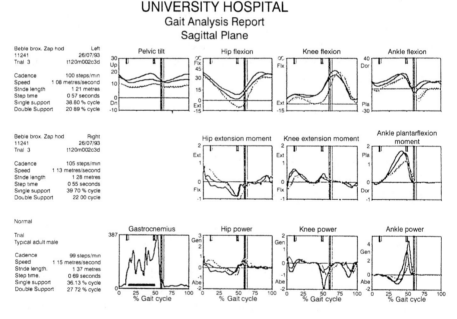

FIGURE 7.1   An example of the "Gait Cycle Window," VICON system. Representation by means of graphical user interface is compact and clearly laid-out, it can be "moved" in space and time, and the user can quickly determine particular gait phases, by means of foot-switch signals (lower part of the figure).

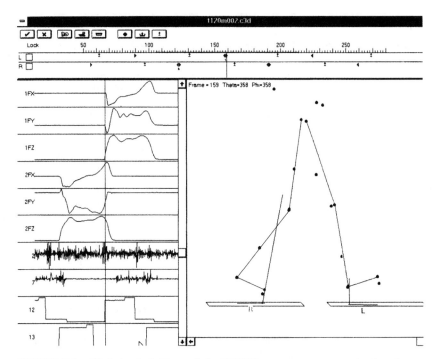

FIGURE 7.2   "Gait Analysis Report," the VICON system, as an example. Angular moments in the sagittal plane, powers, and EMG data are encompassed. The user is able to define desired data choice and data representation.

encompasses the measured angles in a sagittal plane and corresponding calculated moments and powers. Typical results in a normal population (from the user data base) are plotted in green, while data on the left or right lower extremity of a subject are plotted in red or blue. Results are comprehensive. If the VICON VX or the VICON 370, Ethernet LAN, DEC Pathworks, or Windows® for WorkGroups are used, data in the 3D kinematic and analog databases can be accessed easily.

Gage and Koop[8,17] explained clinical application and argument importance of biomechanical gait analysis in children with cerebral paralysis. Due to partial loss of control and muscle weakness, the locomotor process is disturbed in many ways in these individuals. Precise biomechanical measurements in these individuals may contribute significantly to their evaluation, in all phases of treatment and in planning (surgical) procedures.

The SELSPOT MULTILab Motion Analysis System is built around the SELSPOT II system. It enables comprehensive measurements, data processing, and presentation and is applicable in gait analysis in medicine and for measuring robot kinematics in robotics. The TRACK system (in the Newman Laboratory for Rehabilitation Biomechanics at MIT) includes a NEWTON software system. Besides providing precise research measurements, a transfer of solutions to the Biomotion Laboratory at Massachusetts General Hospital in Boston was provided aimed at clinical application. A transfer to the Kinematics Laboratory of the NeuroMuscular

TABLE 7.1
An Elaborated Strategy of the Clinical Application of Gait Analysis (Gait Diagnostic Chart)

Observed Abnormality	Possible Causes	Biomechanical and Neuromuscular Diagnostic Evidence
Foot slap at heel contact	Below normal dorsiflexor activity at heel contact	Below normal tibialis anterior EMG or dorsiflexor moment at heel contact
Forefoot or flatfoot initial contact	(a) Hyperactive plantar-flexor activity in late swing	(a) Above normal plantar-flexor EMG in late swing
	(b) Structural limitation in ankle range	(b) Decreased dorsiflexion range of motion
	(c) Short step-length	(c) See (a), (b), (c), and (d) immediately below
Short step-length	(a) Weak push-off prior to swing	(a) Below normal plantar-flexor moment or power generation (A2) or EMG during push-off
	(b) Weak hip flexors at toe-off and early swing	(b) Below normal hip flexor moment or power or EMG during late push-off and early swing
	(c) Excessive deceleration of leg in late swing	(c) Above normal hamstring EMG or knee flexor moment or power absorption (K4) late in swing
	(d) Above normal contralateral hip extensor activity during contralateral stance	(d) Hyperactivity in EMG of contralateral hip extensors
Stiff-legged weight bearing	(a) Above normal extensor activity at the ankle, knee, or hip early in stance[a]	(a) Above normal EMG activity or moments in hip extensors, knee extensor, or plantar-flexors early in stance
Stance phase with flexed, but rigid knee	(a) Above normal extensor activity in early and middle stance at the ankle and hip, but with reduced knee extensor activity	(a) Above normal EMG activity or moments in hip extensors and plantar-flexors in early and middle stance

Research Center at Boston University, as well as to the University of Bologna, Italy, was also provided.

Based on a 20-year experience in measuring locomotion, Winter[230] described efforts to find clinically useful information for various kinds of gait pathologies: limb amputations, artificial joints, cerebral paralysis, hemiplegia, etc. Table 7.1 provides

TABLE 7.1 (continued)
An Elaborated Strategy of the Clinical Application of Gait Analysis (Gait Diagnostic Chart)

Observed Abnormality	Possible Causes	Biomechanical and Neuromuscular Diagnostic Evidence
Weak push-off accompanied by observable pull-off	(a) Weak plantar-flexor activity at push-off; normal or above normal hip flexor activity during late push-off and early swing	(a) Below normal plantar-flexor EMG, moment or power (A2) during push-off; normal or above normal hip flexor EMG or moment or power during late push-off and early swing
Hip hiking in swing (with or without circumduction of lower limb)	(a) Weak hip, knee, or ankle dorsiflexor activity during swing	(a) Below normal tibialis anterior EMG or hip or knee flexors during swing
	(b) Overactive extensor synergy during swing	(b) Above normal hip or knee extensor EMG or moment during swing
Trendelenburg gait	(a) Weak hip adductors	(a) Below normal EMG in hip abductors: gluteus medius and minimus, tensor fasciae latae
	(b) Overactive hip abductors	(b) Above normal EMG in hip adductors, adductor longus, magnus and bravis, and gracilis

[a] There may be below normal extensor forces at one joint, but only in the presence of abnormally high extensor forces at one or both of the other joints.

From Winter, D. A. 1991. *Gangbildanalyse—Stand der Messtechnik und Bedeutung für die Orthopädie-Technik,* U. Boenick, M. Nader, and Mainka, Eds. Duderstadt: Mecke Druck, 266–277. With permission.

his general strategy for clinical application of gait analysis. Anomalies in gait observed are connected, through possible causes, with biomechanical and neuromuscular findings.

Although contemporary locomotion measurement systems do not provide diagnostics, they do, however, provide differential diagnostics and an evaluation of locomotive problems. Conditions necessary for a measurement method are satisfactorily fulfilled (Section 3.3), while the realization of 3D kinetics certainly gives a new, clinically original dimension. Going back to the 1970s, a time before PCs, we are reminded of the fear of some authors that locomotion measurement procedures could be a purpose unto themselves.[91] The reason for this was probably that there was neither a standardized research methodology of movement analysis in the true sense of the word, nor were clinical applications developed. The backdrop to this was the capital work by Braune and Fischer and Bernstein, characterized by a sophisticated and detailed approach, without the aid of computers. This could have been demoralizing. In time, measurement of human locomotion developed as did general biomechanics at all levels. The globalization of the field was marked by the First World Congress, held in 1990 in San Diego. A similar book by Dainty and Norman ("Standardizing Biomechanical Testing in Sport," 1987) was already a precursor of the revolution in personal computers. The reason for this is not only measurement, but a significant shortening in the time for processing and presentation. The pioneers of the inverse dynamic approach in biomechanics, Braune and Fischer, needed 12 hours to provide a set of experiments and as much as 3 months to analyze measurement data (!) (Ladin,[82] Section 1.2). What was stated more that a decade ago is still true. There is no neuro-musculo-skeletal system diagnostics based on biocybernetic principles.[231] Judging by some recent solutions[46] and resulting commercial products,[232] this might not be true much longer. Multidisciplinary efforts in gait analysis result in clinically applicable products as described in Craik and Oatis.[233] In Europe, a project for the standardization of measurement procedures of locomotion in rehabilitation has been completed, CAMARC (Computer Aided Movement Analysis in the Rehabilitation Context).

7.2. DEVELOPMENT OF METHODOLOGY OF MARKER-FREE KINEMATIC MEASUREMENTS

Insight into engineering solutions for measurement kinematics may lead us to conclude that one of the fundamental problems, not solved satisfactorily at present, is how to determine characteristic body points as accurately as possible and mark them adequately, with the goal of measuring their kinematics and conducting the inverse dynamic procedure. In this context, solutions with CCD matrix sensor-based video cameras, a standard consumer electronics article, have proven to be greatly accessible, offering good possibilities and forcing the high-speed photography method to become obsolete. In clinical locomotion measurements, noninvasiveness is recognized as a basic requirement, i.e., little intrusion on the subject and less burden to the operator. All of this led to development of a system which would measure kinematics using video cameras without requiring markers to be worn, while enabling automatic extraction of kinematic quantities of interest. Therefore, a body segment is identified (upper leg, lower leg, thorax) as a measurement object whose kinematics, then, serve to determine quantities relevant for the estimation of forces and moments. (This is also Furnee's projection.)

After years of clinical experience with the ENOCH system, based on SELSPOT (Section 4.2.1.3), Lanshammar began to develop a method which used the video camera as a kinematic sensor device. The focus was on combining ground reaction force measurements (Kistler platform) and lower extremity kinematics during walking, not whole-body kinematics. Hence, there is no inverse dynamic approach and influence of inertia is neglected, making the approach primarily suitable to studying gait. There are several stages in the development of such a system, whereby video record is combined with ground reaction force, superimposed in the form of a vector diagram, and subsequent off-line digitization of kinematic data is provided (VIFOR and VIFDIG systems). In a more recent method including a software algorithm, a noninvasive estimation of the hip joint center is provided.[234] The first step was a marker-free method, based on a video camera, which estimates center of rotation of one rigid segment only.[235] This method was compared to the equivalent method using markers.[236] Further development included estimation of hip joint rotation, at which point the method was compared to *in vivo* radiological findings in humans.[237] Persson et al.[234] presented the general strategy and validation of the method.

One starts with a one-segment planar representation of a rigid body allowing rotations around a fixed point. This is a crude model, only a first approximation of a lower extremity, and the rotation axis is fixed. In each recorded image (a point in time), the global coordinate system is defined and a local coordinate system is calculated. The center of rotation is then defined as the only fixed point both in the global and the local coordinate system in all images. Then the model is enlarged to include two segments[234] (Figure 7.3.) Experiments were provided with a rigid prosthetic leg model and on healthy subjects. A 50-Hz PAL video camera was used for recording, shutter speed 1/1000, and records were stored on VHS video tape. Three subjects

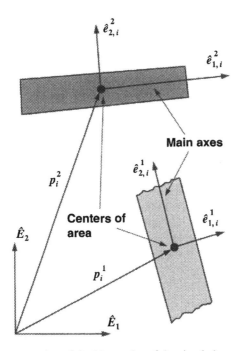

FIGURE 7.3   2D representation of the kinematics of the simple human leg model in a noninvasive marker-free automated measurement method. (From Pershon, T., Lanshammar, H., and Medved, V. 1995. *Comput. Methods Prog. Biomed.* 46:217–224. With permission from Elsevier Science.)

dressed in a black suit participated. Successive video images were digitized off-line (640 $\times$ 480, 6-bit scale of gray) by using a frame-grabber card. In processing digitized images. Global Lab and MatLab software were used. The APAS system was also used. Based on results of comparison to radiological measurements, the accuracy attained was satisfactory and this method was a good starting point for development of noncontact, noninvasive kinematic measurements, aimed for use in clinics.

References

1. Cappozzo, A. 1984. Gait analysis methodology. *Hum. Movement Sci.* 3:27–50.
2. Pierrynowski, M.R. 1988. Computer analysis of locomotion, in *Microcomputers in Physiology—A Practical Approach* (P.J. Fraser, Ed.). Oxford: IRL Press, 179–192.
3. Brand, R.A. 1992. Assessment of musculoskeletal disorders by locomotion analysis: a critical historical and epistemological review, in *Proceedings of the Symposium on Biolocomotion: A Century of Research Using Moving Pictures* (V. Tosi and A. Cappozzo, Eds.). Formia.
4. Messenger, N. 1991. Clinical application of gait analysis: results of a survey in the UK, in *Gangbildanalyse — Stand der Messtechnik und Bedeutung für die Orthopädie-Technik* (U. Boenick, M. Nader, and C. Mainka, Eds.). Duderstadt: Mecke Druck, 191–197.
5. Aiona, M.D. 1996. Human motion analysis: a method of outcome assessment in selective dorsal rhizotomy, in *Human Motion Analysis* (G.F. Harris and P.A. Smith, Eds.). TAB-IEEE Press Book Series, New York: IEEE Press, 303–317.
6. Davis, R.B. and De Luca, P.A. 1996. Clinical gait analysis: current methods and future directions, in *Human Motion Analysis* (G.F. Harris and P.A. Smith, Eds.). TAB-IEEE Press Book Series, New York: IEEE Press, 17–42.
7. Gage, J.R. 1991a. *Gait Analysis in Cerebral Palsy.* Oxford: MacKeith Press.
8. Gage, J.R. 1991b. *Recent Advances in the Treatment of Cerebral Palsy.* The Standard, Oxford Metrics, VICON.
9. Harris, G.F. and Wertsch, J.J. 1996. Human motion analysis—historical perspective and introduction, in *Human Motion Analysis* (G.F. Harris and P.A. Smith, Eds.). TAB-IEEE Press Book Series, New York: IEEE Press, 1–14.
10. Smith, P.A., Harris, G.F., and Abu-Faraj, Z.O. 1996. Biomechanical evaluation of the planovalgus foot in cerebral palsy, in *Human Motion Analysis* (G.F. Harris and P.A. Smith, Eds.). TAB-IEEE Press Book Series, New York: IEEE Press, 370–386.
11. Vaughan, C.L., Damiano, D.L., and Abel, H.F. 1997. Gait of normal children and those with cerebral palsy, in *Three-Dimensional Analysis of Human Locomotion* (P. Allard, A. Coppozzo, A. Lundberg, and C. Vaughan, Eds.). New York: John Wiley & Sons, 335–361.
12. Whittle, M.W. 1995. Musculoskeletal applications of three-dimensional analysis, in *Three-Dimensional Analysis of Human Movement* (P. Allard, I.A.F. Stokes, and J.-P. Blanchi, Eds.). Champaign, IL: Human Kinetics, 295–309.
13. Alexander, I.J., Chao, E.Y.S., and Johnson, K.A. 1990. The assessment of dynamic foot-to-ground forces and plantar pressure distribution: a review of the evolution of current techniques and clinical applications. *Foot Ankle* 11:152–167.
14. Bernstein, N. 1934. The techniques of the study of movements, in *Textbook of the Physiology of Work* (G. Conradi, V. Farfel, and A. Slonim, Eds.). Moscow, *Human Motor Actions — Bernstein Reassessed,* Chapter I (H.T.A. Whiting, Ed.) Amsterdam: Elsevier/North-Holland, 1984, 1–26.
15. Ferrigno, G., Borghese, N.A., and Pedotti, A. 1990. Pattern recognition in 3D automatic human motion analysis. *ISPRS J. Photogramm. Remote Sensing,* 45:227–246.
16. Furnée, E.H. 1989. TV/Computer Motion Analysis Systems: The First Two Decades. Dissertation. Delft University of Technology.

17. Gage, J.R. and Koop, S.E. 1995. Clinical gait analysis: application to management of cerebral palsy, in *Three-Dimensional Analysis of Human Movement* (P. Allard, I.A.F. Stokes, and J.-P. Blanchi, Eds.). Champaign, IL: Human Kinetics, 349–362.
18. Harry, J.D. 1987. Early designs of the myograph. *Med. Instrum.* 21:278–282.
19. Krag, M.H. 1985. Quantitative techniques for analysis of gait. *Automedica* 6:85–97.
20. Lexicon. Jugoslavenski Leksikografski Zavod. 1974 (in Croatian).
21. Marey, E.J. 1895. *Movement.* New York: Appleton (Reprint Edition 1972 by Arno Press, Inc.)
22. McMahon, T.A. 1984. *Muscles, Reflexes, and Locomotion.* Princeton, NJ: Princeton University Press.
23. Nigg, B. and Herzog, W., Eds. 1994. *Biomechanics of the Musculo-Skeletal System.* New York: John Wiley & Sons.
24. Nikolić, V. and Hudec, M. 1988. *Biomechanics—Principles and Elements.* Zagreb: Školska Knjiga (in Croatian).
25. Nilsson, J.E. 1990. On the Adaptation to Speed and Mode of Progression in Human Locomotion. Dissertation. Stockholm: Karolinska Institute and University College of Physical Education.
26. Ranu, H.S. 1991. From the guest editor. *IEEE Eng. Med. Biol. Mag.* 10:12 (Special Issue: *Robotics & Biomechanics*).
27. Seireg, A. and Arvikar, R. 1989. *Biomechanical Analysis of the Musculoskeletal Structure for Medicine and Sports.* New York: Hemisphere.
28. Woltring, H.J. 1977. Marey revisited: prospect and retrospect, in Measurement and Control of Human Movement. Dissertation. Nijmegen: Katholieke Universiteit, 175–199.
29. Woltring, H.J. 1984. On methodology in the study of human movement, in *Human Motor Actions — Bernstein Reassessed,* Chapter Ib (H.T.A. Whiting, Ed.). Amsterdam: Elsevier/ North-Holland, 35–73.
30. Woltring, H.J. 1987. Data acquisition and processing systems in functional movement analysis. *Minerva Orthop. Traumatol.* 38:703–716.
31. Woltring, H.J. 1992. One hundred years of photogrammetry in biolocomotion, in *Proceedings of the Symposium on Biolocomotion: A Century of Research Using Moving Pictures* (V. Tosi and A. Cappozzo, Eds.). Formia.
32. Yanof, H.M. 1965. *Biomedical Electronics.* Philadelphia: F.A. Davis.
33. Berme, N. and Cappozzo, A. 1990. Dynamical analysis, in *Biomechanics of Human Movement: Applications in Rehabilitation, Sports and Ergonomics* (N. Berme and A. Cappozzo, Eds.). Worthington, OH: Bertec Corporation, 103–107.
34. Berme, N., Heydinger, G., and Cappozzo, A. 1987. Calculation of loads transmitted at the anatomical joints, in *Biomechanics of Engineering — Modelling, Simulation, Control* (A. Morecki, Eds.). CISM Udine Courses and Lectures No. 291, Berlin: Springer Verlag, 89–131.
35. Cappozzo, A. 1985. Experimental techniques, data acquisition and reduction, in *Biomechanics of Normal and Pathological Human Articulating Joints* (N. Berme, A.E. Engin, and K.M. Correia da Silva, Eds.). The Hague, The Netherlands: Martinus Nijhoff, 53–81.
36. Brown, G.A., Tello, R.J., Rowell, D., and Mann, R.W. 1987. *Determination of Body Segment Inertial Properties.* RESNA 10th Annual Conference, San Jose, CA, 299–301.
37. Kaufman, K.R. and Sutherland, D.H. 1966. Future trends in human motion analysis, in *Human Motion Analysis* (G.F. Harris and P.A. Smith, Eds.). TAB-IEEE Press Book Series, New York: IEEE Press, 187–215.
38. Davis, R.B., De Luca, P.A., and Õunpuu, S. 1995. Analysis of gait, in *The Biomedical Engineering Handbook* (J.D. Bronzino, Ed.). Boca Raton, FL: CRC Press, 381–390.

39. Hatze, H. 1980. Neuromusculoskeletal control systems modeling — a critical survey of recent developments. *IEEE Trans. Autom. Control.* 25:375–385.
40. Potkonjak, V. 1989. *Robotics.* Beograd: Naučna Knjiga (in Croatian).
41. Heimer, S., Medved, V., and Špirelja, A. 1985. Influence of postural stability on sport shooting performance. *Kineziologija* 17:119–122. (Proceedings: Second Scientific Congress in Sport Shooting — Medicine Psychology Pedagogy in Sport Shooting. Osijek, 1985, 47–56.)
42. Hanavan, E.P. 1964. A Mathematical Model of the Human Body (AMRL-TR-64–102). Wright-Patterson Air Force Base, OH: Aerospace Medical Research Laboratories (NTIS AD-608463).
43. Minns, R.J. 1981. Forces at the knee joint: anatomical consideration. *J. Biomech.* 14:633–643.
44. Kaufman, K.R. and An, K.-N. 1995. Joint-articulating surface motion, in *The Biomedical Engineering Handbook* (J.D. Bronzino, Ed.). Boca Raton, FL: CRC Press, 304–332.
45. Hatze, H. 1981. *Myocybernetic Control Models of Skeletal Muscle — Characteristics and Applications.* Pretoria: Gutenberg.
46. Delp, S.L., Loan, P., Hoy, M.G., Zajac, F.E., Topp, E.L., and Rosen, J.M. 1990. An interactive graphics-based model of the lower extremity to study orthopaedic surgical procedures. *IEEE Trans. Biomed. Eng.* 37(8):757–767.
47. Guyton, A.C. 1987. *Basic Neuroscience — Anatomy and Physiology.* Philadelphia: W.B. Saunders.
48. Keros, P., Posinovec, M., Pećina, M., and Novoselac, M. 1968. *Functional Anatomy of the Locomotor System.* Zagreb: Medicinska Knjiga (in Croatian).
49. Bouisset, S. 1973. EMG and muscle force in normal motor activities, in *New Developments in EMG and Clinical Neurophysiology, Vol. 1* (J.E. Desmedt, Ed.). Basel: S. Karger, 547–583.
50. Waterland, J.C. 1968. Integration of movement, in *Biomechanics I — 1st International Seminar, Zürich, 1967* (J. Wartenweiler, E. Jokl, and M. Hebbelnick, Eds.). Basel: S. Karger, 178–187.
51. Basmajian, J.V. and De Luca, C. 1985. *Muscles Alive — Their Functions Revealed by Electromyography.* Baltimore: Williams & Wilkins.
52. Dietz, V. and Noth, J. 1980. Elektromyographische und kinesiologische Analyse von Sportleistungen, in *Die Belastungstoleranz des Bewegungsapparates* (H. Cotta, H. Krahl, and K. Steinbruck, Eds.). Stuttgart: Georg Thieme Verlag, 22–34.
53. Leonard, C.T., Hirschfeld, H., and Forssberg, H. 1991. The development of independent walking in children with cerebral palsy. *Dev. Med. Child Neurol.* 33:567–577.
54. Forssberg, H., Hirschfeld H., and Stokes, V.P. 1991. Development of human locomotor mechanisms, in *Neurobiological Basis of Human Locomotion* (M. Shimamura, S. Grillner, and V.R. Edgerton, Eds.). Tokyo: Japan Scientific Societies Press, 259–273.
55. Vodovnik, L. 1985. *Neurocybernetics.* Ljubljana: Fakulteta za Elektrotehniko (in Slovene).
56. Rieder, H. 1980. Die variable Verfügbarkeit sportlicher Bewegungsabläufe als Lernprozess, in *Die Belastungstoleranz des Bewegungsapparates* (H. Cotta, H. Krahl, and K. Steinbrück, Eds.). Stuttgart: Georg Thieme Verlag.
57. Hildreth, E.C. and Hollerbach, J.M. 1987. Artificial intelligence: computational approach to vision and motor control, in *Handbook of Physiology. Section 1: The Nervous System* (V.B. Mountcastle, Eds.). Bethesda, MD: American Physiological Society, 605–641.

58. Winters, J.M. 1995. Concepts in neuromuscular modeling, in *Three-Dimensional Analysis of Human Movement* (P. Allard, I.A.F. Stokes, and J.-P. Blanchi, Eds.). Champaign, IL: Human Kinetics, 257–292.

59. Beulke, H. 1978. Kritische Aspekte zur Elektrostimulation als Trainingsmittel. *Leistungssport* 3:224–235.

60. Eyzaguirre, C. and Fidone, S.J. 1975. *Physiology of the Nervous System. 2nd Edition.* Chicago: Year Book Medical Publishers, 167–179.

61. Katz, B. 1979. *Nerv, Muskel und Synapse.* Stuttgart: Georg Thieme Verlag.

62. Hatze, H. 1979. A teleological explanation of Weber's law and the motor unit size law. *Bul. Math. Biol.* 41:407–425.

63. Hill, A.V. 1938. *First and Last Experiments on Muscle Mechanics.* New York: Oxford University Press.

64. Cavanagh, P.R. 1988. On "muscle action" vs. "muscle contraction." *J. Biomech.* 21, 1:69

65. Zajac, F.E. 1989. Muscle and tendon: properties, models, scaling, and application to biomechanics and motor control. *CRC Crit. Rev. Biomed. Eng.* 17:359–411.

66. Chen, A. 1991. Neural Networks: Computing's Next Frontier. *Microcomputer Solutions (A Publication of Intel Corporation),* March/April: 3–6.

67. Souček, B. and Souček, M. 1988. *Neural and Massively Parallel Computers.* New York: John Wiley & Sons.

68. Medved, V. 1984. On analog-to-digital conversion of some physiological and kinesiological signals. *Naučno-Tehnički Pregled.* 34:35–41 (in Croatian).

69. Bowen, B.A. and Brown, W.R. 1982. *VLSI Systems Design for Digital Signal Processing. Volume I: Signal Processing and Signal Processors.* Englewood Cliffs, NJ: Prentice Hall.

70. Stojević, Z., Milevoj, V., Gold, H., Medved, V., Stepinac, J., Trinajstić, D., and Ulaga, I. 1983. Universal digital device for industrial measurements. 11th Exposition of Discoveries, Technical Improvements and Innovations, INOVA '83, exponat no. 134. Zagreb (in Croatian).

71. Medved, V. 1985a. A method to compute a periodogram for an EMG signal. *Naučno-Tehnički Pregled.* 35:32–42 (in Croatian).

72. Medved, V. 1985b. Technical Solution of Multichannel Subsystem for Signal Conditioning and Acquisition Connected to a Microcomputer. Technical note. Zagreb: Faculty of Physical Education (in Croatian).

73. Medved, V. 1988. Comparative Analysis of Bioelectric and Biomechanical Features of Lower Extremity Muscles in Sports Activities. Ph.D. thesis, University of Zagreb, Faculty of Electrical Engineering (in Croatian).

74. Atha, J. 1984. Current techniques for measuring motion. *Appl. Ergon.* 15:245–257.

75. Dainty, D.A. and Norman, R.W. 1987. *Standardizing Biomechanical Testing in Sport.* Champaign, IL: Human Kinetics.

76. Mejovšek, M. and Medved, V. 1982. The Activity of the Biomechanics Laboratory. Zagreb: Faculty of Physical Education (in Croatian).

77. Mann, R., W., Rowell, D., Dalrymple, G., Conati, F., Tetewsky, A., Ottenheimer, D., and Antonsson, E. 1981. Precise, rapid, automatic 3-D position and orientation tracking of multiple moving bodies, in *Proceedings of the VIIIth International Congress of Biomechanics,* Nagoya.

78. Chao, E.Y. 1978. Experimental methods for biomechanical measurements of joint kinematics, in *CRC Handbook of Engineering in Medicine and Biology. Section B: Instruments and Measurements. Volume I* (B.N. Feinberg and D.G. Fleming, Eds.). Boca Raton, FL: CRC Press 385–411.

79. Medved, V. 1980. Initial Evaluation of CARS-UBC Electrogoniometer System. Technical Note. Cambridge, MA: Harvard University, Biomechanics Laboratory, Department of Applied Sciences.

80. Bajd, T., Kljajić, M., Trnkoczy, A., and Stanić, U. 1974. Electrogoniometric measurement of step length. *Scand. J. Rehab. Med.* 6:78–80.

81. Chao, E.Y.S. and An, K.N. 1982. Perspectives in measurement and modelling of musculoskeletal joint dynamics, in *Biomechanics: Principles and Applications* (R. Huiskes, D. Van Campen, and J. De Wijn, Eds.). The Hague, The Netherlands: Martinus Nijhoff Publishers, 1–18.

82. Ladin, Z. 1995. Three-dimensional instrumentation, in *Three-Dimensional Analysis of Human Movement* (P. Allard, I.A.F. Stokes, and J.-P. Blanchi, Eds.). Champaign, IL: Human Kinetics, 3–17.

83. Woltring, H.J. and Huiskes, R. 1990. Measurement of body segment motion, in *Biomechanics of Human Movement: Applications in Rehabilitation, Sports and Ergonomics* (N. Berme and A. Cappozzo, Eds.). Worthington, OH: Bertec Corporation, 108–127.

84. Abdel-Aziz, Y.I. and Karara, H.M. 1971. Direct linear transformation from comparator coordinates into object-space coordinates in close-range photogrammetry, in *Proceedings of the ASP/UI Symposium "Close-Range-Photogrammetry,"* Urbana, IL, American Society of Photogrammetry, Falls Church, VA, 1–18.

85. Woltring, H.J. 1975a. Calibration and measurement in 3 dimensional monitoring of human motion by optoelectronic means. I. Preliminaries and theoretical aspects, in *Biotelemetry* (H.P. Kimmich, Ed.). Basel: S. Karger, 169–196.

86. Sutherland, D.H. and Hagy, J.L. 1972. Measurement of gait movements from motion picture film. *J. Bone Jt. Surg.* 54-A(4), 787–797.

87. Saunders, M., Inman, V., and Eberhart, H.D. 1953. The major determinants in normal and pathological gait. *J. Bone Jt. Surg.* 35-A(3), 543–558.

88. Cappozzo, A. and Paul, J. 1997. Instrumental observation of human movement: historical development, in *Three-Dimensional Analysis of Human Locomotion* (P. Allard, A. Cappozzo, A. Lundberg, and C. Vaughan, Eds.). New York: John Wiley & Sons, 1–25.

89. Forssberg, H. and Hirshfeld, H. 1988. Phasic modulation of postural activation patterns during human walking, *Prog. Brain Res.* 76: 221–227.

90. Greaves, J.O.B. 1995. Instrumentation in video-based three-dimensional systems, in *Three-Dimensional Analysis of Human Movement* (P. Allard, I.A.F. Stokes, and J.-P. Blanchi, Eds.). Champaign, IL: Human Kinetics, 41–54.

91. Cavanagh, P.R. 1976. Recent advances in instrumentation and methodology of biomechanical studies, in *Biomechanics* V-B, 399–411.

92. Furnée, E.H. 1991. Opto-electronic movement measurement systems: aspects of data acquisition, signal processing and performance, in *Gangbildanalyse — Stand der Messtechnik und Bedeutung für die Orthopädie-Technik* (U. Boenick, M. Nader, and C. Mainka, Eds.). Duderstadt: Mecke Druck, 112–129.

93. Furnée, E.H. 1990. PRIMAS: Real-time image-based motion measurement system, in *SPIE Vol. 1356, Image-Based Motion Measurement* (Proceedings Mini-Symposium on Image-Based Motion Measurement, First World Congress on Biomechanics, San Diego, CA, J.S. Walton, Ed.). Bellingham, WA: Society of Photo-Optical Engineers, 56–62.

94. Jarett, M.O., Andrews, B.J., and Paul, J.P. 1976. A television/computer system for the analysis of human locomotion. IERE Golden Jubilee Conference on the Applications of Electronics in Medicine, University of Southampton, England, April 6–8, 1976. *IERE Conference Proc.* Vol. 34., 357–370.

95. Winter, D.A., Greenlaw, R.K., and Hobson, D.A. 1972. Television-computer analysis of kinematics of human gait. *Comput. Biomed. Res.* 5:498–504.

96. Herrmann, J. 1991. Primas precision motion analysis system — a short description, in *Gangbildanalyse — Stand der Messtechnik und Bedeutung für die Orthopädie-Technik* (U. Boenick, M. Nader, and C. Mainka, Eds.). Duderstadt: Mecke Druck, 60–65.

97. Ferrigno, G. and Pedotti, A. 1985. ELITE: a digital dedicated hardware system for movement analysis via real time TV signal processing. *IEEE Trans. Biomed. Eng.* 32:943–950.

98. Macleod, A., Morris, J.R.W., and Lyster, M. 1990. Highly accurate video coordinate generation for automatic 3D trajectory calculation, in SPIE Vol. 1395, *Close-Range Photogrammetry Meets Machine Vision* (A. Gruen and E. Baltsavias, Eds.). Bellingham, WA: SPIE — The International Society for Optical Engineering, 12–17.

99. Whittaker T.D. 1991. Recent software developments for biomechanical assessment via Expert Vision, in *Gangbildanalyse — Stand der Messtechnik und Bedeutung für die Orthopädie-Technik* (U. Boenick, M. Nader, and C. Mainka, Eds.). Duderstadt: Mecke Druck, 66–70.

100. Borghese. N.A. and Ferrigno, G. 1990. An algorithm for 3-D automatic movement detection by means of standard TV cameras. *IEEE Trans. Biomed. Eng.* 37, 12:1221–1225.

101. Greiff, H. and Theysohn, H. 1991. Das Hentschel-System HSG 84.330, projektiert an der Universität Munster, in *Gangbildanalyse — Stand der Messtechnik und Bedeutung für die Orthopädie-Technik* (U. Boenick, M. Nader, and C. Mainka, Eds.). Duderstadt: Mecke Druck, 71–77.

102. Emmons, R.B. 1967. The lateral photoeffect. *Solid State Electron.* 10: 505–506.

103. Woltring, H.J. 1974. New possibilities for human motion studies by real-time light spot position measurement. *Biotelemetry* 1:132–146.

104. Woltring, H.J. 1975b. Single- and dual-axis lateral photodetectors of rectangular shape. *IEEE Trans. Electron Devices* 22:581–590.

105. Krzystek, P. 1990. Real-time photogrammetry with lateral-effect photodiodes. State of the art and recent investigations, in SPIE Vol. 1395, *Close-Range Photogrammetry Meets Machine Vision* (A. Gruen and E. Baltsavias, Eds.). Bellingham, WA: SPIE — The International Society for Optical Engineering, 30–37.

106. Westermark, J. 1991. The Selspot Multilab-System, in *Gangbildanalyse — Stand der Messtechnik und Bedeutung für die Orthopädie-Technik* (U. Boenick, M. Nader, and C. Mainka, Eds.). Duderstadt: Mecke Druck, 19–22.

107. Gustafsson, L. and Lanshammar, H. 1977. ENOCH — An Integrated System for Measurement and Analysis of Human Gait. Dissertation. Uppsala: Uppsala University, Institute of Technology.

108. Stokes, V.P. 1984. A method for obtaining the 3 D kinematics of the pelvis and thorax during locomotion. *Hum. Movement Sci.* 3:77–94.

109. Stokes, V.P., Andersson, C., and Forssberg, H. 1989. Rotational and translational movement features of the pelvis and thorax during adult human locomotion. *J. Biomech.* 22(1):43–50.

110. Crouch, D.G., Kehl, L., and Krist, J.R. 1990. Optotrak — At last a system with resolution of ten microns, in SPIE Vol. 1356, *Image-Based Motion Measurement* (Proceedings Mini-Symposium on Image-Based Motion Measurement, First World Congress on Biomechanics, San Diego, CA, J.S. Walton, Ed.). Bellingham, WA: Society of Photo-Optical Engineers, 53 (Abstract).

111. Krist, J., Melluish, M., Kehl, L., and Crouch, D. 1991. Technical description of the Optotrak 3D Motion Measurement System, in *Gangbildanalyse—Stand der Messtechnik und Bedeutung für die Orthopädie-Technik* (U. Boenick, M. Nader, and C. Mainka, Eds.). Duderstadt: Mecke Druck, 23–39.

112. Macellari, V. 1983. CoSTEL: a computer peripheral remote sensing device for 3-dimensional monitoring of human motion. *Med. Biol. Eng. Comput.* 21:311–318.

113. Bianchi, G., Gazzani, F., and Macellari, V. 1990. The Costel system for human motion measurement and analysis, in SPIE Vol. 1356, *Image-Based Motion Measurement* (Proceedings Mini-Symposium on Image-Based Motion Measurement, First World Congress on Biomechanics, San Diego, CA, J.S. Walton, Ed.). Bellingham, WA: Society of Photo-Optical Engineers, 38–51.

114. Michelson, D. 1990. CODA: Evolution and application, in *Image-Based Motion Measurement*, SPIE Vol. 1356, J.S. Walton, Ed. Bellington, WA: International Society of Photo-Optical Engineers, 26–37.

115. Zebris, commercial material.

116. MacReflex, commercial material.

117. Stussi, E. and Müller, R. 1991. Vergleichende Bewertung kommerziell erhältlicher 3D-Kinematik-Systeme für die Gangbildanalyse, in *Gangbildanalyse—Stand der Messtechnik und Bedeutung für die Orthopädie-Technik* (U. Boenick, M. Nader, and C. Mainka, Eds.). Duderstadt: Mecke Druck, 86–97.

118. Koff, D.G. 1995. Joint kinematics: camera-based systems, in *Gait Analysis: Theory and Application* (R.L. Craik and C.A. Oatis, Eds.). St. Louis: C.V. Mosby, 183–204.

119. Lamoreux, L.W. 1995. Coping with soft tissue movement in human motion analysis, in *Human Motion Analysis* (G.F. Harris and P.A. Smith, Eds.). TAB-IEEE Press Book Series, New York: IEEE Press, 43–70.

120. Divljaković, V. 1978. Optoelectronic Methods for Determination of the Object's Spatial Position. M.Sc. thesis. University of Zagreb, Faculty of Electrical Engineering (in Croatian).

121. Divljaković, V. 1982. The Influence of Opto-electrical Transformation on the Accuracy of Aerial Measurement by Optoelectronic Planimetry. Ph.D. thesis. University of Zagreb, Faculty of Electrical Engineering (in Croatian).

122. Skala, K. 1983. Analysis of Reflection Detectability in Wide Angle Noncoherent Optical Illumination. Ph.D. thesis. University of Zagreb, Faculty of Electrical Engineering (in Croatian).

123. Pedotti, A. and Ferrigno, G. 1995. Optoelectronic-based systems, in *Three-Dimensional Analysis of Human Movement* (P. Allard, I.A.F. Stokes, and J.-P. Blanchi, Eds.). Champaign, IL: Human Kinetics, 57–77.

124. Lanshammar, H. 1982a. On practical evaluation of differentiation techniques for human gait analysis. *J. Biomech.* 15:99–105.

125. Lanshammar, H. 1982b. On precision limits for derivatives numerically calculated from noisy data. *J. Biomech.* 15:459–470.

126. Lee, M.H. 1989. *Intelligent Robotics*. Milton Keynes: Open University Press.

127. Marr, D. 1982. *Vision*. New York: W.H. Freeman.

128. Chao, E.Y.S. 1990. Goniometry, accelerometry, and other methods, in *Biomechanics of Human Movement: Applications in Rehabilitation, Sports and Ergonomics* (N. Berme and A. Cappozzo, Eds.). Worthington, OH: Bertec Corporation, 130–139.

129. Ladin, Z. and Wu, G. 1991. Combining position and acceleration measurements for joint force estimation. *J. Biomech.* 24(12):1173–1187.

130. Woltring, H.J. 1985. On optimal smoothing and derivative estimation from noisy displacement data in biomechanics. *Hum. Movement Sci.* 4:229–245.

131. Wood, G.A. 1982. Data smoothing and differentiation procedures in biomechanics. *Exercise Sport Sci. Rev.* 10:308–362.

132. Wiener, N. 1950. *Extrapolation, Interpolation and Smoothing of Stationary Time Series.* New York: John Wiley & Sons.

133. Allard, P., Blanchi, J.-P., and Aissaoui, R. 1995. Bases of three-dimensional reconstruction, in *Three-Dimensional Analysis of Human Movement* (P. Allard, I.A.F. Stokes, and J.-P. Blanchi, Eds.). Champaign, IL: Human Kinetics, 19–40.

134. Pezzack, J.C., Norman, R.W., and Winter, D.A. 1977. An assessment of derivative determining techniques used for motion analysis. *J. Biomech.* 10:377–382.

135. Wood, G.A. and Jennings, L.S. 1979. On the use of spline functions for data smoothing. *J. Biomech.* 12:477–479.

136. Woltring, H.J. 1995. Smoothing and differentiation techniques applied to 3-D data, in *Three-Dimensional Analysis of Human Movement* (P. Allard, I.A.F. Stokes, and J.-P. Blanchi, Eds.). Champaign, IL: Human Kinetics, 79–99.

137. Smith, G. 1989. Padding point extrapolation techniques for the Butterworth digital filter. *J. Biomech.* 22(8/9):967–971.

138. D'Amico, M.D. and Ferrigno, G. 1992. Comparison between the more recent techniques for smoothing and derivative assessment in biomechanics. *Med. Biol. Eng. Comput.* 30:193–204.

139. Mottet, D., Bardy, B.G., and Athenes, S. 1994. A note on data smoothing for movement analysis: the relevance of a nonlinear method. *J. Motor Behav.* 26(1):51–55.

140. Winter, D.A., Eng, J.J., and Ishac, M.G. 1996. Three-dimensional moments, powers, and work in normal gait. Implications for clinical assessments, in *Human Motion Analysis* (G.F. Harris, and P.A. Smith, Eds.). TAB-IEEE Press Book Series, New York: IEEE Press, 71–83.

141. Komi, P.V. 1990. Relevance of *in vivo* force measurements to human biomechanics. *J. Biomech.* 23(Suppl. 1):23–34.

142. Medved, V. 1987. Instrumenti, in *Sportska Medicina* (R. Medved, Ed.). Zagreb: Jumena, 789–810.

143. Božičević, J. 1982. *Foundations of Automatics 2. Measurement Transducers and Measurement, Second Edition.* Zagreb: Školska Knjiga (in Croatian).

144. Berme, N. 1990. Load transducers, in *Biomechanics of Human Movement: Applications in Rehabilitation, Sports and Ergonomics* (N. Berme and A. Cappozzo, Eds.). Worthington, OH: Bertec Corporation, 141–149.

145. Barnes, S.Z. and Berme, N. 1995. Measurement of kinetic parameters technology, in *Gait Analysis: Theory and Application* (R.L. Craik and C.A. Oatis, Eds.). St. Louis: C.V. Mosby, 239–251.

146. Cromwell, L., Weibel, F.J., and Pfeiffer, E.A. 1980. *Biomedical Instrumentation and Measurements.* Englewood Cliffs, NJ: Prentice-Hall.

147. De Marre, D. and Michaels D. 1983. *Bioelectronic Measurements.* Englewood Cliffs, N.J.: Prentice-Hall.

148. Gautschi, G.H. 1978. Piezoelectric multicomponent force transducers and measuring systems. Transducer '78 Conference.

149. Geddes, L.A. and Baker, L.E. 1968. *Principles of Applied Biomedical Instrumentation.* New York: John Wiley & Sons.

150. Nilsson, J., Thorstensson, A., and Halbertsma, J. 1985. Changes in leg movements and muscle activity with speed of locomotion and mode of progression in humans. *Acta Physiol. Scand.* 123:457–475.

151. Milinković, G. 1988. Comparison of Ground Reaction Force in Healthy and Disturbed Gait — Lower Leg Amputations. B.Ed. thesis. University of Zagreb, Faculty of Physical Education (in Croatian).

152. Hallett, M., Stanhope, S.J., Thomas, S.L., and Massaquoi, S. 1991. Pathophysiology of posture and gait in cerebellar ataxia, in *Neurobiological Basis of Human Locomotion* (M. Shimamura, S. Grillner, and V.R. Edgerton, Eds.). Tokyo: Japan Scientific Societies Press, 275–283.

153. Cavanagh, P.R. and Lafortune, M.A. 1980. Ground reaction forces in distance running. *J. Biomech.* 13:397–406.

154. Milanović, D., Heimer, S., Medved, V., Mišigoj-Duraković, M., and Fattorini, I. 1989. Possibilities of application of test results in programming top athletes' training. *Basketball Med. Periodical* 4(1):3–8.

155. Janković, V. 1997. Personal communication.

156. Clarke, H.H. 1953. *Application of Measurement to Health and Physical Education.* New York: Prentice Hall.

157. Zmajić, H. 1988. Relations Between Take-Off Force Biomechanical Parameters and Success in High Jump. B.Ed. thesis. Zagreb: Faculty of Physical Education (in Croatian).

158. Medved, V., Wagner, I., and Živčić, K. 1985. Biomechanical Testing of Take-Off Abilities in Sports Gymnastics. Zagreb: Faculty of Physical Education (in Croatian).

159. Medved, V. and Wagner, I. 1987. Biomechanical Testing of Take-Off Abilities in Sports Gymnastics II. Zagreb: Faculty of Physical Education (in Croatian).

160. Medved, V. 1989. Bioelectric and kinetic locomotion diagnostics. *Elektrotehnika* 32, 6:335–343 (in Croatian).

161. George, G.S. 1980. *Biomechanics of Women's Gymnastics.* Englewood Cliffs, NJ: Prentice Hall.

162. Pedotti, A. 1982. Functional evaluation and recovery in patients with motor disabilities, in *Uses of Computers in Aiding the Disabled — IFIP IMIA* (J. Raviv, Ed.). Amsterdam: North-Holland, 53–71.

163. Crenna, P. and Frigo, C. 1985. Monitoring gait by a vector diagram technique in spastic patients, in *Clinical Neurophysiology in Spasticity* (P.J. Delwaide and R.R. Young, Eds.). Amsterdam: Elsevier 109–124.

164. Santambrogio, G.C. 1989. Procedure for quantitative comparison of ground reaction data. *IEEE Trans. Biomed. Eng.* 36:247.

165. Hasan, S.S., Robin, D.W., and Shiavi, R.G. 1992. Drugs and postural sway. Quantifying balance as a tool to measure drug effects. *IEEE Eng. Med. Biol.* 11:35–41.

166. Shimba, T. 1984. An estimation of center of gravity from force platform data. *J. Biomech.* 17:53–60.

167. Šimunjak, B. 1994. Stabilometry in the Evaluation of Capabilities and Training of Iceskaters. M.Sc. thesis. Zagreb: Medical School (in Croatian).

168. Hasan, S.S. and Robin, D. 1992. Developing force platform measures of postural sway as indicators of drug effects. Bulletin: STABAPP-0592.

169. Schumann, T., Redfern, M.S., Furman, J.M., El-Jaroudi, A., and Chaparro, L.F. 1995. Time-frequency analysis of postural sway. *J. Biomech.* 28:603–607.

170. Medved, V. and Heimer, S. 1993. A new balance measurement system: some analytical and empirical considerations. *Period. Biolog.* 95:101–104.

171. Hirschfeld, H. 1992. On the Integration of Posture, Locomotion and Voluntary Movement in Humans: Normal and Impaired Development. Dissertation. Stockholm: Karolinska Institute.

172. Cobb, J. and Claremont, D.J. 1995. Transducers for foot pressure measurement: survey of recent developments. *Med. Biol. Eng. Comput.* 33:525–532.

173. Lord, M. 1981. Foot pressure measurement: a review of methodology. *J. Biomed. Eng.* 3:91–99.

174. Pedotti, A., Assente, R., and Fusi, G. 1984. Multisensor piezoelectric polymer insole for pedobarography. *Ferroelectrics* 60:163–174.

175. Peruchon, E., Jullian, J.-M., and Rabischong, P. 1989. Wearable unrestraining footprint analysis system. Applications to human gait study. *Med. Biol. Eng. Comput.* 27:557–565.

176. Cavanagh, P.R. and Ae, M. 1980. A technique for the display of pressure distribution beneath the foot. *J. Biomech.* 13:69–75.

177. Anon. 1991. A pressure mapping system for gait analysis. *Sensors.* 21–23.

178. Awbrey B.J., Etskovitz, R.B., Richlen, C.A., Podoloff, R.M., and Benjamin, M.H. 1991. A new flexible sensor to evaluate dynamic in-shoe plantar pressure during gait and its use to measure the effect of shoe accomodation devices in severe metatarsaglia, in Combined Meeting of the Orthopaedic Research Societies of U.S.A., Japan, and Canada, Banff, Alberta.

179. Lord, M., Hosein, R., and Williams, R.B. 1992. Method for in-shoe shear stress measurement. *J. Biomed. Eng.* 14, 181–186.

180. Skalak, R. and Chen, S., Eds. 1987. *Handbook of Bioengineering.* New York: McGraw-Hill.

181. Enoka, R. 1994. *Neuromechanical Basis of Kinesiology.* Champaign, IL: Human Kinetics.

182. Hardt, D.E. and Mann, R.W. 1980. A five body-three dimensional dynamic analysis of walking. *J. Biomech.* 13:455–457.

183. Margaria, R. 1976. *Biomechanics and Energetics of Muscular Exercise.* Oxford: Clarendon Press.

184. Winter, D.A., 1979. *Biomechanics of Human Movement.* New York: John Wiley & Sons.

185. Winter, D.A., Eng, J.J., and Ishac, M.G. 1995. A review of kinetic parameters in human walking, in *Gait Analysis: Theory and Application* (R.L. Craik and C.A. Oatis, Eds.). St. Louis: C.V. Mosby, 252–270.

186. Medved, R. et al. 1987. *Sports Medicine.* Zagreb: Jumena (in Croatian).

187. Geddes, L.A. 1995a. Historical perspectives, Section 3, Recording of action potentials, in *The Biomedical Engineering Handbook* (J.D. Bronzino, Ed.). Boca Raton, FL: CRC Press, 1367–1377.

188. Geddes, L.A. 1995b. Historical perspectives, Section 4, Electromyography, in *The Biomedical Engineering Handbook* (J.D. Bronzino, Ed.). Boca Raton, FL: CRC Press, 2144–2151.

189. Herzog, W., Guimaraes, A.C.S., and Zhang, Y.T. 1994. EMG, in *Biomechanics of the Musculo-Skeletal System* (B.M. Nigg and W. Herzog, Eds.). New York: John Wiley & Sons, 308–336.

190. Clarys, J.P. 1994. Electrology and localized electrization revisited. *J. Electromyogr. Kinesiol.* 4:5–14.

191. Henneberg, K.-A. 1995. Principles of electromyography, in *The Biomedical Engineering Handbook* (J.D. Bronzino, Ed.). Boca Raton, FL: CRC Press, 191–199.

192. Rozendal, R.H. and Meijer, O.G. 1982. Human kinesiological electromyography: some methodological problems. *Hum. Movement Sci.* 1:7–26.

193. Landowne, D. 1987. Neural conduction, in *Handbook of Bioengineering* (R. Skalak and S. Chen, Ed.). New York: McGraw-Hill, 28.1–28.17.

194. De Luca, C.J. 1979. Physiology and mathematics of myoelectric signals. *IEEE Trans. Biomed. Eng.* 26(6):313–326.

195. De Luca, C.J. 1984. Myoelectrical manifestations on localized muscular fatigue in humans. *CRC Rev. Biomed. Eng.* 11:251–279.

196. Loeb, G.E. and Gans, C. 1986. *Electromyography for Experimentalists*. Chicago: The University of Chicago Press.

197. Cifrek, M. 1997. Myoelectric Signal Analysis During Dynamic Fatigue. Ph.D. thesis. University of Zagreb, Faculty of Electrical Engineering and Computing (in Croatian).

198. De Luca, C.J. 1995. *Surface Electromyography: Detection and Recording*. Boston: Boston University, NeuroMuscular Research Center.

199. De Luca, C.J. and Knaflitz, M. 1992. *Surface Electromyography: What's New?* Boston: Neuromuscular Research Center.

200. O'Connel, A.L. and Gardner, E.T. 1964. The use of electromyography in kinesiological research. *Res. Q.* 34:166–184.

201. Basmajian, J.V. 1968. The present status of electromyographic kinesiology, in Biomechanics I, 1st Int. Seminar, Zürich, 1967 (J. Wartenweiler, E. Jokl, and M. Hebbelinck, Eds.). Basel: S. Karger, 110–122.

202. Winter, D.A., Rau, G., Kadefors, R., Broman, H., and De Luca, C.J. 1980. Units, Terms and Standards in the Reporting of EMG Research. Report by the Ad Hoc Commitee of the ISEK.

203. Neuman, M.R. 1995b. Biopotential electrodes, in *The Biomedical Engineering Handbook* (J.D. Bronzino, Ed.). Boca Raton, FL: CRC Press, 745–757.

204. Šantić, A. 1988. *Electronic Instrumentation*. Zagreb: Školska Knjiga (in Croatian).

205. Šantić. A. 1995. *Biomedical Electronics*. Zagreb: Školska Knjiga (in Croatian).

206. Viitasalo, J.T., Saukkonen, S., and Komi, P.V. 1980. Reproducibility of measurements of selected neuromuscular performance variables in man. *Electromyogr. Clin. Neurophysiol.* 20:487–501.

207. Nagel, J.H. 1995. Biopotential amplifiers, in *The Biomedical Engineering Handbook* (J.D. Bronzino, Ed.). Boca Raton, FL: CRC Press, 1185–1195.

208. Winter, D.A., Fuglevand, A.J., and Archer, S.E. 1994. Cross talk in surface electromyography: theoretical and practical estimates. *J. Electromyogr. Kinesiol.* 4(1):15–26.

209. Welkowitz, W., Deutsch, S., and Akay, M. 1992. *Biomedical Instruments: Theory and Design*. New York: Academic Press.

210. Sepulveda, F., Wells, D.M., and Vaughn, C.L. 1993. A neural network representation of electromyography and joint dynamics in human gait. *J. Biomech.* 26(2):101–109.

211. Ericson M., O., Nisell, R., and Ekholm, J. 1986. Quantified electromyography of lower-limb muscles during level walking. *Scand. J. Rehab. Med.* 18:159–163.

212. Perry, J. and Bekey, G.A. 1981. EMG-Force relationships in skeletal muscle. *Crit. Rev. Biomed. Eng.* 7:1–22.

213. Lawrence, J.H. and De Luca, C.J. 1983. Myoelectric signal versus force relationship in different human muscles. *J. Appl. Physiol. Respir. Environ. Exercise Physiol.* 54(6):1653–1659.

214. Duchene, J. and Goubel, F. 1993. Surface electromyogram during voluntary contraction: Processing tools and relation to physiological events. *Crit. Rev. Biomed. Eng.* 21(4):313–397.

215. Dowling J.J. 1997. The use of electromyography for the noninvasive prediction of muscle force — current issues. *Sports Med.* 24(2): 82–96.

216. Gandy et al. 1980. Acquisition and analysis of electromyographic data associated with dynamic movements of the arm. *Med. Biol. Eng. Comput.* 57.

217. Chen, J.J. and Shiavi, R. 1990. Temporal feature extraction and clustering analysis of electromyographic linear envelopes in gait studies. *IEEE Trans. Biomed. Eng.* 37(3):295–302.

218. Shiavi, R. and Griffin, P. 1981. Representing and clustering electromyographic gait patterns with multivariate techniques. *Med. Biol. Eng. Comput.* 19:605–611.

219. Shiavi, R., Bourne, J., and Holland, A. 1986. Automated extraction of activity features in linear envelopes of locomotor electromyographic patterns. *IEEE Trans. Biomed. Eng.* 33(6):594–600.

220. Berme, N. 1980. Control and movement of lower limbs, in *Biomechanics of Motion* (A. Morecki, Ed.). New York: Springer, 185–217.

221. Milner, M., Basmajian, J.V., and Quandbury, A.O. 1971. Multifactorial analysis of walking by electromyography and computer. *Am. J. Phys. Med.* 50(5):235–258.

222. Pedotti, A. 1977. A study of motor coordination and neuromuscular activities in human locomotion. *Biol. Cybernet.* 26:53–62.

223. Perry, J., 1996. The role of EMG in gait analysis, in *Human Motion Analysis* (G.F. Harris and P.A. Smith, Eds.). TAB-IEEE Press Book Series, New York: IEEE Press, 255–267.

224. Thorstensson, A., Carlson, H., Zomlefer, M.R., and Nilsson, J. 1982. Lumbar back muscle activity in relation to trunk movements during locomotion in man. *Acta Physiol. Scand.* 116:13–20.

225. Medved, V. and Tonković, S. 1991a. Method to evaluate the skill level in fast locomotion through myoelectric and kinetic signal analysis. *Med. Biol. Eng. Comput.* 29:406–412.

226. Medved, V. and Tonković, S. 1991b. Locomotion diagnostics: some neuromuscular and robotic aspects. *IEEE Eng. Med. Biol. Mag.* 10(2):23–28 (Special Issue: *Robotics & Biomechanics*)

227. Medved, V., Tonković, S., and Cifrek, M. 1995. Simple neuro-mechanical measure of the locomotor skill: an example of backward somersault. *Med. Progr. Technol.* 21:77–84.

228. Schwartz, M. 1975. *Signal Processing: Discrete Spectral Analysis, Detection, and Estimation.* New York: McGraw-Hill.

229. Spiegel, M.R. 1992. *Theory and Problems in Statistics. Schaum's Outline Series.* New York: McGraw-Hill.

230. Winter, D.A. 1991. Integrated gait analysis as a diagnostic tool in orthopaedics and prosthetics, in *Gangbildanalyse — Stand der Messtechnik und Bedeutung für die Orthopädie-Technik* (U. Boenick, M. Nader, and C. Mainka, Eds.). Duderstadt: Mecke Druck, 266–277.

231. Medved, V. and Medved, Ve. 1984. Trend of development of medical diagnostic systems based on bioelectric signal processing. Proceedings of the Symposium Biokibernetika '84. Skopje III.13–III.20 (in Croatian).

232. MusculoGraphics, commercial material.

233. Craik, R.L. and Oatis, C.A. 1995. *Gait Analysis: Theory and Application.* St. Louis: C.V. Mosby.

234. Persson, T., Lanshammar, H., and Medved, V. 1995. A marker-free method to estimate hip joint centre of rotation by video image processing. *Comput. Methods Prog. Biomed.* 46:217–224.

235. Sahlen, T., Lanshammar, H., and Hansson, K. 1993. Marker-free estimation of hip joint position, in *Proceedings of the International Society of Biomechanics XIVth Congress, Paris.*

236. Lanshammar, H., Persson, T., and Medved, V. 1994a. Comparison between a marker-based and a marker-free method to estimate centre of rotation using video image analysis, in *Second World Congress of Biomechanics, Volume II* (L. Blankevoort and J.G.M. Koolos, Eds.). Amsterdam, 375.

237. Lanshammar, H., Persson, T., and Medved, V. 1994b. Validation of a marker-free method to estimate hip joint position by video image processing, in *World Congress on Medical Physics and Biomedical Engineering, Rio de Janeiro.*

Appendix 1

ISB Recommendations for Standardization in the Reporting of Kinematic Data*

Since 1990, the Standardization and Terminology Committee of the International Society of Biomechanics has been working toward a recommendation for standardization in the reporting of kinematic data. The recommendations, which are a result of those efforts (including broad input from members of the Society), are intended as a guide to the presentation of kinematic data in refereed publications and other materials. It is hoped that some uniformity in presentation will make publications easier to read and allow for the more straightforward comparison of data sets from different investigators. It is not intended to restrict individual investigators in the manner in which they collect or process their data. Rather, it could be viewed as a "output filter" applied to a variety of data formats to provide uniformity in the final product.

The ISB is cognizant of the various attempts at standardization that are being pursued by other organizations—such as CAMARC in Europe, the Clinical Gait Laboratory Group in the U.S., and the efforts of individual professional groups such as the Scoliosis Research Society. Where possible, we have sought unanimity with these groups, but on issues where the members of our society expressed strong opinions, we have—at times—stated a contrary view. One example in point is the use of center of mass-based segmental reference frames. Since such reference frames are needed for conventional dynamic analysis, we make the recommendation that such frames should be routinely used. We anticipate that extension to the present document in the future will include recommendations for joint coordinate systems and the definition of anatomical landmarks for locating other segmental reference frames.

The committee recognizes that standardization of the description of movement at individual joints is best left to those who are intimately involved in the study of those joints, and we have therefore appointed subcommittees for various joints to provide recommendations. Groups are currently active for the ankle, hand, shoulder, spine, temporomandibular joints, whole body, and wrist. Members with interests and

---

* Reprinted from *J. Biomechanics*, Vol. 28(10), Wu, G. and Cavanaugh, P.R., *ISB Recommendations for Standardization in the Reporting of Data*, 1257–1261, 1995. Copyright © 1995, with permission from Elsevier Science.

expertise in other joints are being actively sought. The initial recommendations of some of these groups have already been published in the *ISB Newsletter,* and once these recommendation have been discussed by the membership, a subsequent document on joint coordinate systems will be published.

The present recommendations are presented as a framework on which future progress can be based. We are grateful to former members of the Standardization and Terminology Committee (notably Professors John Paul, David Winter, and Don Grieve) and to the many ISB members who commented on earlier drafts of this recommendation. The present recommendations owe much to the work of Sommer (1991).

PART 1: DEFINITION OF A GLOBAL REFERENCE FRAME

Need: A global reference frame with the direction of the global axes being consistent, no matter which activities or subjects are being studied or which investigator is conducting the experiment

Notation: X, Y, Z

Recommendation: A right-handed orthogonal triad fixed in the ground with the $+ Y$ axis upward and parallel with the field of gravity, X and Z axes in a plane that is perpendicular to the Y axis

Notes: (a) Where there is clear direction of travel or work (as is the case for level gait), the $+ X$ axis is defined as the direction of travel or work (see Figure 1).

(b) In case of locomotion on inclined planes, the Y axis will remain vertical and the X and Z axes will be in the same horizontal plane.

(c) Where there is no clear direction of travel or work (as is the case for insect flight), the $+ X$ axis should be defined by the investigator.

(d) In tasks such as exercise in zero gravity, the $+ X$ axis should be defined according to some arbitrary but visible surface in the environment and in the direction that is meaningful to the task.

(e) We acknowledge that there may be situations where non-Cartesian axes are more appropriate to the task being studied (for example, cylindrical coordinates are useful for the study of asymmetric manual exertion). Since the majority of studies use a Cartesian approach, it will be left to individual investigators to devise systems for the reporting of more specialist situations.

(f) The directions of the $X, Y,$ and Z axes have been chosen so that for those conducting two-dimensional studies, X, Y will lie in a sagittal plane. This will be consistent with the three-dimensional convention.

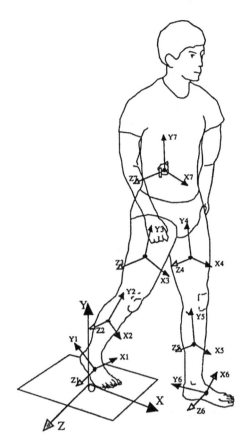

FIGURE 1   Conventions for global reference frame and segmental local center of mass reference frame.

PART 2: DEFINITION OF SEGMENTAL LOCAL CENTER OF MASS REFERENCE FRAMES

Need:	A coherent frame to describe segment post (position and orientation) with respect to the global frame
Notation:	X_i, Y_i, Z_i
Recommendation:	A series of right-handed orthogonal triads fixed at the segmental centers of mass with the $+ X_i$ axis anterior, $+ Y_i$ axis proximal, and $+ Z_i$ being defined by a right-hand rule
Notes:	(a) In general, the anterior-posterior, proximal-distal, and medial-lateral directions are defined in relation to the standard anatomical position.
	(b) The use of right-hand reference frames for both left- and right-body segments implies that for the segments on the right side of the body, the $+ Z_i$ is

pointing laterally, and for the segments on the left side of the body, the $+ Z_i$ is pointing medially (Figure 1). As a result, the positive movements about the X_i and Y_i axes of a segment on the left side of the body will have opposite effects of movements of similar sign on the right side of the body (Figure 2). This difference should be accounted for by describing the movements in their anatomical terms in any presentation of the data.

PART 3: GLOBAL DISPLACEMENT

Need: Specification of displacement of a segment with respect to the global reference frame
Notation: x_i, y_i, z_i
Recommendation: Report the coordinates of the origins of the segmental local center of mass reference frames with respect to the global origin in meters. The position of the local origin will be represented by the first column of the 4 × 4 matrix in the local to global transformation matrix $[T_{lg}]$ (see below).

PART 4: GLOBAL ORIENTATION

Need: Specification of orientation of a segment with respect to the global reference frame
Notation: $\alpha_i, \beta_i, \gamma_i$
Recommendation: A standard *ZYX* decomposition (Sommer, 1991) of the 3 × 3 rotation submatrix of the 4 × 4 matrix will be used to define the local to global transformation matrix $[T_{lg}]$:

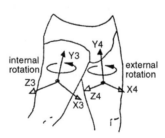

FIGURE 2 The same rotations about segmental local center of mass reference frames produce anatomically different motions on the left and right sides of the body.

$$[T_{1g}] = \begin{bmatrix} 1 & 0 & 0 & 0 \\ x_i & C_{11i} & C_{21i} & C_{31i} \\ y_i & C_{12i} & C_{22i} & C_{32i} \\ z_i & C_{13i} & C_{23i} & C_{33i} \end{bmatrix}$$

where C_{11i}, C_{21i}, C_{31i} are the direction cosines of the local X_i axis with respect to the global X, Y, and Z axes, respectively.

Note: If α_i, β_i, γ_i is an ordered series of rotations about Z_i, Y_i, and X_i axes, respectively, then

$$[T_{1g}] = \begin{bmatrix} 1 & 0 & 0 & 0 \\ x_i & c\,\alpha_i c\,\beta_i & c\,\alpha_i s\,\beta_i s\,\gamma_i - s\,\alpha_i c\,\gamma_i & c\,\alpha_i s\,\beta_i c\,\gamma_i + s\,\alpha_i s\,\gamma_i \\ y_i & s\,\alpha_i c\,\beta_i & s\,\alpha_i s\,\beta_i s\,\gamma_i + c\,\alpha_i c\,\gamma_i & s\,\alpha_i s\,\beta_i c\,\gamma_i - c\,\alpha_i s\,\gamma_i \\ z_i & -s\,\beta_i & c\,\beta_i s\,\gamma_i & c\,\beta_i c\,\gamma_i \end{bmatrix}$$

where $s\,\alpha_i = \sin(\alpha_i)$ and $c\,\alpha_i = \cos(\alpha_i)$.
The individual angles can be found as follows:
$\beta_i = -\arcsin(C_{31i})$,
$\alpha_i = \arcsin(C_{21i}/\cos(\beta_i))$,
$\alpha_i = \arccos(C_{11i}/\cos(\beta_i))$,
$\gamma_i = \arcsin(C_{32i}/\cos(\beta_i))$,
$\gamma_i = \arccos(C_{33i}/\cos(\beta_i))$.

PART 5: RELATIVE ORIENTATION

Need: A frame (or system) to express the relative orientation of the body segments with respect to each other

Notation: α: rotation about one axis of the proximal segment's local reference frame

                 γ: rotation about one axis of the distal segment's local reference frame

                 β: rotation about the floating axis

Recommendation: A joint coordinate system (which might better be called a joint rotation convention) is defined for each joint individually. This system allows rotations about axes which can be anatomically meaningful at the sacrifice of establishing a reference frame with nonorthogonal axes. As long as forces and moments are not resolved along these nonorthogonal axes, this does not present a problem. This approach allows the preservation of an important linkage with clinical medicine where the use of independent paired rotations (ab/ad, internal/external, etc.) is common usage. The most well-known examples of such systems are

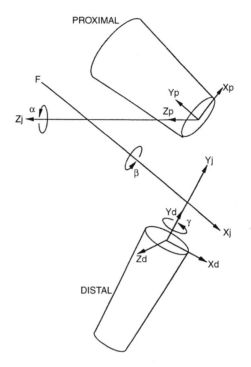

FIGURE 3   A joint coordinate system for the knee joint.

Notes:

those developed for the knee by Grood and Suntay (1983) and Chao (1986) (Figure 3). Two body fixed axes are established relative to anatomical landmarks, one in each body on opposing sides of the joint. The third axis, called the floating axis, is defined as being perpendicular to each of the two body fixed axes.

(a) The orientation of the axes in the local reference frames of the proximal and distal segments must be clearly specified.

(b) The choice of the location of the origins drastically affects the distraction displacement terms.

(c) The Euler angle set in Part 4 (global orientation) should match the angle decomposition for the joints as closely as possible.

REFERENCES

Chao, E. Y. S. (1986) Biomechanics of human gait. In *Frontiers in Biomechanics*. (Edited by Schmid-Schonbein, G. W., Woo. S. L.-Y., and Zweifach, B. W.), Springer, New York.
Grood, E. S. and Suntay, W. J. (1983) A joint coordinate system for the clinical description of three-dimensional motions: application to the knee. *J. Biomechanical Eng.* 105, 136–144.

Sommer, H. J., III (1991) *Primer on 3-D Kinematics.* Handout to the American Society of Biomechanics Meeting. Tempe, AZ, U.S.

EDITORIAL COMMENT

The recommendations on terminology for the reporting of kinematic data represent a thoughtful approach arising from a committee of the International Society of Biomechanics. The material was developed over some years and involved a number of individuals. The recommendations are published without peer review, since they arose from a committee of one of our participating organizations and since they represent a compilation of commonly accepted practices.

R. A. Brand
Editor

Appendix 2

Standards for Reporting EMG Data*

The following information must be supplied when reporting EMG data.

Electrodes:
Reports on *surface recording* of EMG should include:

- electrode material (e.g., Al/AgCl, etc.)
- electrode geometry (discs, bars, rectangular, etc.)
- size (e.g., diameter, radius, width × length)
- use of gel or paste, alcohol applied to cleanse skin, skin abrasion, shaving of hair, etc.
- interelectrode distance
- electrode location, orientation over muscle with respect to tendons, motor point and fibers direction

Intramuscular wire electrodes should be described by:

- wire material (e.g., stainless steel, etc.)
- if single- or multistrand
- insulation material
- length of exposed tip
- method of insertion (e.g., hypodermic needle, etc.)
- depth of insertion
- if single or bipolar wire
- location of insertion in the muscle
- interelectrode distance
- type of ground electrode used, location

Needle electrodes and their application should be described according to standard clinical protocol. The use of nonstandard needle electrodes should be fully described

* Reprinted from *J. Electromyogr. Kinesiol.*, 4, 1994. Copyright © 1994, with permission from Elsevier Science.

and include material, size (gauge), number and size of conductive contact points at the tip, depth of insertion and accurate location in the muscle.

Amplification:
Amplifiers should be described by the following:

- if single, differential, double differential, etc.
- input impedance
- Common Mode Rejection Ratio (CMRR)
- signal-to-noise ratio
- actual gain range used

Filtering of the raw EMG should be specified by:

- low and/or high pass filters
- filter types (e.g., Butterworth, Chebyshev, etc.)
- low and/or high pass cut-off frequencies

Since the power density spectra of the EMG contains most of its power in the frequency range of 5–500 Hz at the extremes, the journal will not accept reports in which surface EMG was filtered above 10 Hz as a low cut-off, and below 350 Hz as the high cut-off; e.g., 10–350 Hz is preferred for *surface* recordings. Filtering in the band of 10–150 Hz or 50–350 Hz, for example, is not acceptable as portions of the signal's power above 150 Hz and below 50 Hz are eliminated. This should be kept in mind when designing a study's protocol. Exceptions will be made only in rare cases that carry full scientific justification.

Intramuscular recording should be made with the appropriate increase of the high frequency cut-off to a minimum 450 Hz. A bandpass filter of 10–450 Hz is therefore required.

Needle recording should have a bandwidth of 10–1500 Hz.

Rectification: A note should be made if full or half-wave rectification was carried out.

EMG Processing: There are several methods of EMG processing.

Smoothing the signal with a low pass filter of a given time constant (normally 50–250 ms) is best described as "smoothing with a low-pass filter of x ms." Alternately, one can describe it as a "linear envelope" or "the mean absolute value," while giving time constant type and order of the low-pass filter used. Designating the EMG subjected to this procedure as the "integrated EMG" (IEMG) is incorrect (see below).

Another acceptable method is determination of the *"root mean square"* or RMS. Authors should include the time (period) over which the average RMS was calculated.

Integrated EMG is sometimes reported, but the signal is actually integrated over time, rather than just smoothed. Such procedure allows observation of the accumulated EMG activity over time and should be presented with information as to whether time or voltage was used to reset the integrator and at what threshold.

Power density spectra presentation of the EMG should include:

- time epoch used for each calculation segment
- type of window used prior to taking the Fast Fourier transform (FFT) (e.g., Hamming, Hanning, Tukey, etc.)
- taking the algorithm (e.g., FFT)
- number of zero padding applied in the epoch and the resultant resolution
- equation used to calculate the Median Frequency (MDF), Mean Frequency (MNF), etc.
- the muscle length or fixed joint angle at the time of recording

Other processing techniques, especially novel techniques, are encouraged if accompanied by full scientific description.

Sampling EMG into the Computer:
Computer processing of the EMG is encouraged if authors observe these important factors:

1. It is advisable that the raw EMG (e.g., after differential amplification and bandpass filtering) be stored in the computer before further analysis in case modification of the protocol is required in the future. In this case, the minimal acceptable sampling rate is at least twice the highest frequency cut-off of the bandpass filter, e.g., if a bandpass filter of 10–350 Hz was used, the minimal sampling rate employed to store the signal in the computer should be 700 Hz (350 × 2) and *preferably higher* to improve accuracy and resolution. Sampling rates below twice the highest frequency cut-off will not be accepted.
2. If smoothing with a low-pass filter was performed with hardware prior to sampling and storing data in the computer, the sampling rate could be drastically reduced. Rates of 50–100 Hz are sufficient to introduce smoothed EMG into the computer.
3. It is also advisable that authors consider recording the raw EMG (prior to bandpass filtering) in the computer; in such cases a sampling rate of 2500 Hz or above could be used. Yet, to avoid aliasing of high-frequency noise, bandpass filtering (written in software) in the range prescribed above should be performed prior to any further processing of the signal. This approach allows authors to perform EMG recording with minimal hardware and maximal flexibility. Yet, it may be at the expense of taxing computer memory space and speed.
4. Number of bits, model, and manufacturer of A/D card used to sample data into the computer should be given.

Normalization: In investigations where the force/torque was correlated to the EMG, it is common to normalize the force/torque and its respective EMG, relative to the values at maximal voluntary contraction (MVC). Authors should be aware that obtaining true MVC from subjects requires some preliminary training. Without training, the

MVC could be as much as 20–40% less of that obtained after appropriate training and lead to incorrect conclusions or interpretation of data. The journal, therefore, will not accept reports in which subjects were not properly trained to elicit true MVC.

Normalizing the force/torque with respect to its MVC is commonly performed with MVC as 100% of the force/torque, and other force levels are expressed as the appropriate % of MVC. Similarly, the EMG associated with 100% MVC is designated as 100% and fractions thereof. Both force/torque and EMG normalization should include other relevant information such as joint angle(s) and/or muscle length(s) in isometric contractions, and range of joint angle, muscle length, velocity of shortening/elongation, and load applied for non-isometric contractions.

Normalization of data collected from one experimental condition with respect to other contractile conditions can be performed for comparative purposes and will be accepted by the journal only if full description and justification are given.

In summation, the following information should be provided when normalizing data:

- how subjects were trained to obtain MVC
- joint angle or muscle length
- angles of adjoining joint, e.g., for studies on elbow flexion, the position of the wrist and shoulder joints should be provided
- rate of rise of force
- velocity of shortening/elongation
- changes in muscle length
- ranges of joint angle/muscle length in non-isometric contraction
- load applied in non-isometric contractions

EMG Cross-talk:
Authors should demonstrate that an earnest effort was undertaken to determine that EMG cross-talk from muscles near the muscle of interest did not contaminate the recorded signal. Selecting the appropriate electrode size, interelectrode distance, and location of recordings over the muscle should be carefully planned, especially when working on areas where many narrow muscles are tightly gathered (e.g., forearm) or when working with superficial/thin muscles (e.g., trapezius). The work of Winter (1994) and Fugelvand et al. (1992) should be consulted if doubts exist. Care also should be employed when recording surface EMG from areas with subcutaneous adipose tissue (e.g., abdomen, buttocks, chest, etc.) as it is known that adipose tissue enhances cross-talk (Solomonow et al., 1994).

Index